John Naughton is Vice-President of Wolfson College, Cambridge. He is also the *Observer*'s 'Networker' columnist and a prominent blogger at memex.naughtons.org. His last book was *A Brief History of the Future: The Origins of the Internet* (1999). He lives in Cambridge.

FROM GUTENBERG TO ZUCKERBERG

What You Really Need to Know About the Internet

John Naughton

Quercus

First published in Great Britain in 2012 by Quercus

This paperback edition published in 2012 by
Quercus
55 Baker Street
7th Floor, South Block
W1U 8EW

A CIP catalogue record for this book is available
from the British Library

ISBN PB 978 0 85738 426 3
ISBN EBOOK 978 0 85738 547 5

10 9 8 7 6 5 4 3 2 1

Text designed and typeset by Ellipsis Digital Limited, Glasgow

Printed and bound in Great Britain by Clays Ltd, St Ives plc

For Jasper, who came into the world at the time that this book was conceived.

'But life is short and information endless: nobody has time for everything. In practice we are generally forced to choose between an unduly brief exposition and no exposition at all. Abbreviation is a necessary evil and the abbreviator's business is to make the best of a job which, though intrinsically bad, is still better than nothing. He must learn to simplify, but not to the point of falsification. He must learn to concentrate upon the essentials of a situation, but without ignoring too many of reality's qualifying side issues. In this way he may be able to tell, not indeed the whole truth (for the whole truth about almost any important subject is incompatible with brevity), but considerably more than the dangerous quarter-truths and half-truths which have always been the current coin of thought.'

Aldous Huxley, *Brave New World Revisited*

CONTENTS

PROLOGUE: WHY THIS BOOK?

It all started with a question that I'm frequently asked. Because I once wrote a history of the Internet[1] – and people infer from that that I must have some superior insight into it – they eventually get round to posing a question that's clearly been bugging them for a while. The question is couched in various ways, but always comes down to the same thing: why is the Net such a big deal? The motivation for asking the question varies. Some people are baffled by the Internet, and wonder how it works. Some are awestruck by it, and wonder what it means for our societies and our futures. Some are irritated by it – especially by what they see as the social and behavioural changes that it appears to be driving, or by the challenges that it poses to the established order. Some are enraged by it – especially if they see it as threatening their livelihoods or undermining their professional authority. Some are fearful of it as an alienating, atomizing technology that turns us (and, more importantly, our children) into screen-based addicts, locked into silent communion with remote objects. And so on.

As the old adage puts it, the worst kind of fear is fear of the unknown. So my inference from all this is that the root of the unease is a lack of understanding. This is not as dismissive as it sounds. I don't mean that my interrogators are stupid, or ignorant. Nor are they uninformed. Their problem is not that they are short of information about the Net; *au contraire*, they are awash with conflicting data about it. It's just that they don't know what it all means. They are in the state once described by that great scholar of cyberspace, Manuel Castells, as 'informed bewilderment'.[2]

This book is an attempt to alleviate that bewilderment. It stems from an attempt to explain what the Internet phenomenon means for all of us. I sat down one day and asked myself the question: *what would you really need to know to understand the Internet and its implications?*

It turns out that you don't need to know all that much. You need to understand some history and some basic technological ideas. You need to know a little about the architectural principles that underpin the network's design. You need a different perspective on networking than you will get from the news media. And most of all you need to look at the Internet phenomenon in terms of what it's doing to our media ecosystem – the information environment that influences how human culture evolves. What this amounts to is a smallish number of Big Ideas which will, I hope, increase enlightenment and reduce the bewilderment of which Castells writes so eloquently.

But how many Big Ideas? In thinking about this I was reminded of a celebrated paper by the psychologist George Miller of Princeton University. Its title was 'The Magical Number Seven, Plus or Minus Two: Some Limits on our Capacity for Processing Information' and it was based on a talk he gave on 15 April 1955 to the Eastern Psychological Association in Philadelphia, and published the following year in the journal *Psychological Review*.[3] Miller's objectives when he gave the talk were pretty modest: he simply set out to summarize some earlier psychological experiments which attempted to measure the limits of people's short-term memory. In each case he reported that the effective 'channel capacity' lay between five and nine equally weighted choices. Miller did not draw any firm conclusions from this though, and contented himself by merely conjecturing that 'the recurring sevens might represent something deep and profound or be just coincidence'. And that, he probably thought, was that.

But he underestimated the appetite of popular culture for anything with the word 'magical' in the title. Instead of being known as a mere aggregator of research results, Miller found himself identified as a kind of sage – a discoverer of a profound truth about human nature. 'My problem,' he wrote,

> is that I have been persecuted by an integer. For seven years this number has followed me around, has intruded in my most private data, and has assaulted me from the pages of our most public journals. This number assumes a variety of disguises, being sometimes a little larger and sometimes a little smaller than usual, but never changing so much as to be unrecognizable. The persistence with which this number plagues me is far more than a random accident. There is, to quote a famous senator,[4] a design behind it, some pattern governing its appearances. Either there really is something unusual about the number or else I am suffering from delusions of persecution.

But in fact the basic idea that emerges from the 1956 paper still seems useful. The idea is that our short-term memory can only hold between five and nine 'chunks' of information at any given time – where a chunk is defined as a 'meaningful unit'.

ying to decide how many Big Ideas about the
ld be meaningful for most readers, it seemed
settle for George Miller's magical number – seven
plus or minus two. And from that decision, everything
else followed.

So what are these ideas? There are nine, each of which has a chapter to itself.

1 Take the long view
We're in the midst of a major upheaval in our information environment, and none of us has any real idea of where it will end up. So we need to put it into perspective. As it happens, Johannes Gutenberg triggered a comparable revolution five and a half centuries ago when he introduced printing by moveable type. His invention shaped the world into which the Internet was born. What can we learn from that experience?

2 The Web is not the Net
The World Wide Web is vast and important, but it's not the Internet. Nor is Google the Internet. Nor is Facebook. These are just three of the things that run on the Internet's infrastructure. The network is bigger and more important than any of the applications or services that it supports. It's vital that governments, policy-makers and users understand this because . . .

3 For the Net, disruption is a feature, not a bug
Why has the Internet triggered such an explosion of innovation and creativity? The answer lies deep in its architecture, and in the principles that underpinned its original design. The archi-

tects of the network effectively created a global machine for springing surprises. So disruptive innovation is, in a way, what the Internet was designed to foster. Some of the surprises it has sprung on us have been pleasant, others less so. But the only way we will stop them coming is either to switch off the network, or to cripple it in ways that will staunch the creative flow.

4 Think ecology, not just economics

We need new intellectual frameworks – new tools for thinking – to understand what's going on. Economics is usually the discipline of choice for thinking about major shifts in behaviour and it has its uses in thinking about the Internet, especially in relation to regulation. But because it focuses on markets and the allocation of scarce resources, economics often struggles to explain what happens in cyberspace, which is characterized by an abundance of resources, and a great deal of activity that takes place entirely outside the market system. Ecology – the study of natural systems – provides a useful alternative framework for thinking about what's going on. In particular, it suggests why our emerging media ecosystem might be considerably more productive than the one it is in the process of replacing.

5 Complexity is the new reality

Our new media ecosystem is immeasurably more complex than the one in which most of us were educated and conditioned. Yet complexity is something that we have traditionally tried to ignore or control. Since denial and control are no longer options, we need to tool up for the challenge. In particular, we need to pay attention to how complex systems work, and to how our organizations need to be reshaped to make them cope with the complexity that now confronts them.

6 The network is now the computer

Once upon a time, the computer was 'Big Iron'[5] – a huge expensive mainframe which took up a large room. After that it was a 'minicomputer' the size of a large domestic refrigerator, and after that it morphed into the PC sitting on your desk. But then, imperceptibly, as access to the Internet improved and **bandwidth** increased, the location of the processing power needed to give us the services that we once got from PCs shifted; it moved off our work-tables and desks and onto distant servers based somewhere on the Internet. This is what is called '**cloud computing**', and it has become an indispensable part of online life – with serious implications for privacy, security and the environment.

7 The Web is evolving

The Web was once a fairly static, read-only medium. But that was a long time ago. In fact it's been in a continuous state of evolution ever since it was launched at the beginning of the 1990s – an evolution caricatured by the progression from 'Web 1.0' to 'Web 2.0' and, now, 'Web 3.0'. What's happening is that the Web, like the Internet on which it runs, has itself become a platform for innovation. So, in a sense, when people log onto what we call the World Wide Web, they are increasingly logging into the World Wide Computer. And the process of innovation that drives that shift is accelerating, not slowing.

8 Copyrights and 'copywrongs': or why our intellectual property regime no longer makes sense

Digital technology allows us to make perfect copies of everything – and to distribute them globally. But our laws on intellectual property were framed in an era in which copying was relatively difficult and imperfect, and in which dissemination was expensive and unreliable. We're now on a catastrophic collision course between the capabilities of networked technology and the powerful vested interests which control intellectual property. How did we get into this crazy position? And how might we get out of it?

9 Orwell vs. Huxley: the bookends of our networked future?
Where is all this technical evolution taking us? Whatever happened to techno-Utopia – the fantasy that the Internet would change everything for the better? The strange irony is that our networked futures can be mapped as the space between the visions of two English twentieth-century writers – George Orwell and Aldous Huxley. Orwell thought we would be destroyed by the things we fear. Huxley thought that it's the things we enjoy that would prove our undoing. Who has turned out to be the more perceptive guide to the future of cyberspace?

The first thing to be said about these ideas is that none is particularly profound. When I first published an outline of this book in the *Observer*,[6] the comments from readers made instructive reading. A smallish minority loudly proclaimed that there was nothing new here, that these ideas are obvious and 'old hat' (as that lovely English phrase puts it). 'This article,' said one critic, 'reads as if it is written by an 80 year old for other 80 year olds. Something to talk about at Bingo.'

My guess is that much of the dismissive criticism came from the ranks of the so-called 'technorati'[7] – people who understand this stuff and think about it a lot. A good many other readers who do not have technology backgrounds observed (in emails and face-to-face encounters) that at least some of the ideas were new to them. In saying that, they were confirming what two and a half decades spent thinking, arguing, lecturing and writing about the Net have taught me, namely that our society has become critically dependent on a technology that is poorly understood, not just by its users, but also by people (like government ministers) who are in a position to make decisions about how it should be regulated and controlled.

Why does this matter? It matters to me because my (baby-boomer) generation invented something magical – a network that shrank the world, gave us access to a wealth of information and knowledge that would have been inconceivable to earlier generations and stimulated a wave of disruptive innovations that could enable our future prosperity in hitherto undreamt-of ways. But this invention of ours is not a fixed, immutable object: it's changing and evolving and could be perverted or throttled by the powerful industrial and political forces that it has threatened; or it could be emasculated by clueless legislation and ill-informed regulation. Entrusting the future of the Internet to such forces would be as foolish as giving a delicate watch to a monkey.

Secondly, there's an inevitable arbitrariness about any list of 'key' ideas. I'm absolutely certain that if one consulted other experts and commentators they would produce their own lists of critical concepts. My hope is that there would be some overlap between mine and theirs, but one never knows with these things. All I can do is to try to distil what I've learned in the last couple of decades, and hope that it makes sense to you.

1. TAKE THE LONG VIEW

We're in the throes of a revolution. And the strange thing about living through a revolution is that it's very difficult to see what's going on. Imagine what it must have been like, for example, being a resident of St Petersburg in the early autumn of 1917, in the months before Lenin and the Bolsheviks finally seized power in Russia. It's clear that momentous events are afoot. There are all kinds of conflicting rumours and theories in circulation. But nobody knows how things will pan out. Only with the benefit of hindsight will we get a clear idea of what was going on. But the clarity that hindsight bestows is also misleading, because it understates how confusing things appeared to people at the time.

So it is with us now. We're living through a radical transformation of our communications environment. Since we don't have the benefit of hindsight, we don't really know where it's taking us. And one thing we've learned from the history of communications technology is that people tend to over-estimate the short term impact of new technologies – and to *underestimate* their long-term implications.

We see this all around us at the moment, as would-be *savants*, commentators, writers, consultants and visionaries tout their personal interpretations of What The Internet Means for business, publishing, retailing, education, politics and the Future of Civilization As We Know It. Often these interpretations are compressed into vivid slogans, **memes*** or aphorisms. Information, we are told, 'wants to be free'. The 'Long Tail' is the future of retailing. Google is 'making us stupid'. Print is dead. Texting is destroying grammar. Censorship is impossible (because, in the words of a famous engineer,[1] 'the Internet interprets censorship as damage and routes around it'). Privacy is a quaintly outmoded concept. The online consumer is 'sovereign'. Markets are becoming 'frictionless'. 'Content is king' – or not, as the case may be. The world is 'flat'[2]. And so on.

These kinds of slogans are really just short-term extrapolations from yesterday's or today's experience. They tell us little about where the revolution we're currently living through is ultimately headed. The question is: can we do any better, without falling into the trap of feigning omniscience? Remember that I'm no smarter than you are. And there's no point in pretending that the future is anything other than unknowable.

So here's a radical idea: why not see if there's anything to be learned from history? And here we get a lucky break. Because it just so happens that mankind – or at any rate that portion of it that inhabits Western Europe and territories shaped by it – has lived through an earlier transformation in its communications environment. It was brought about by the invention of printing by moveable type. Over a number of centuries, this technology changed the world – indeed it shaped the cultural environment in which most of us grew up. And the great thing about it from the point of view of this essay is

* Terms in bold are defined in the Glossary at the end of the book.

that we can view it with the benefit of hindsight. We know what happened.

A thought experiment

So let's conduct what the Germans call a *Gedankenexperiment* – a thought experiment. Let's imagine that print is an historical analogy for the Internet – that the Net represents a similar kind of transformation in our communications environment. What would we learn from such an experiment?

We can date the invention of printing by moveable type from 1455, when the first printed Bibles emerged from the press created by Johannes Gutenberg in the German city of Mainz.

Now imagine that the year is 1473 – that's eighteen years after 1455. Imagine further that you're the medieval equivalent of a MORI pollster, standing on the bridge in Mainz with a clipboard (clipslate?) in your hand, stopping pedestrians and requesting permission to ask them a few questions.

Here's Question 4:

> On a scale of 1 to 5, where 1 indicates 'Not at all likely' and 5 indicates 'Very likely', how likely do you think it is that Herr Gutenberg's invention will:
> (a) Undermine the authority of the Catholic Church?
> (b) Trigger a Protestant Reformation?
> (c) Enable the rise of modern science?
> (d) Create entirely new social classes and professions?
> (e) Change our conceptions of 'childhood' as a protected early period in a person's life?

On a scale of 1 to 5! You only have to ask the questions to realize the fatuity of the idea. Printing did indeed have all of these effects, but there was no way that in 1473 anyone in Mainz – or anywhere else for that matter –could have known how profound its impact would be.

Which brings me back to the Internet. I'm writing this in 2011, which is eighteen years since the Web went mainstream and thereby brought the Internet into the consciousness of the general public. If I'm right about the Web bringing about a transformation in our communications environment comparable to that wrought by print, then it's obviously absurd for me (or anyone else) to pretend to know what its long-term impact will be. The honest answer is that we haven't a clue.

That doesn't mean, of course, that we cannot make guesses. And this is where the print analogy is particularly helpful. By looking in more detail at the transformations that printing brought about, we can perhaps get an idea of where we should be looking for the longer-term impact of the Net.

So let's go digging.

The typescript on the wall

It's always difficult to pinpoint the exact date at which a revolution happened. And there are usually arguments about originality and chronology. Thus we now know that the Chinese had invented a form of printing by moveable type (initially using clay type, later with wood blocks) by the end of the first millennium. Wood-block printing onto cloth first came to Europe perhaps a century later and, once methods of manufacturing paper had been introduced, 'block-books' printed on paper appeared from about 1425 onwards.

But for our purposes, Johannes Gutenberg – or to give him his full handle, Johannes Gensfleisch zur Laden zum Gutenberg – is the main man. For such an important figure in our collective history, we know astonishingly little about him.[3] What we do know is that he was born in 1398 in Mainz, then a small city in Germany on the banks of the Rhine, that he came from a prosperous family and was probably apprenticed as a goldsmith. We also know that in 1411 his family decamped suddenly from Mainz, that he may have been a student at the University of Erfurt and that he wound up in Strasbourg in the early 1430s. Beyond that, everything is pretty vague.

From what we do know based on his innovations, however, we can infer that Gutenberg must have been an archetypal geek. That is to say, he must have been obsessive, ingenious, persistent, fanatical, infuriating and mostly broke. The obsessiveness was necessary because in order to achieve the breakthrough for which he is remembered he had to assemble a whole raft of technologies and make them work together. For example, he had to: find a method of making metallic type from an alloy of lead, tin and antimony; produce inks that were based on oil rather than water; invent a method of casting the type in a hand mould; and build a press (derived from the ancient wine-press) which could reliably impress an image of the inked type on a sheet of paper. And last but by no means least, he had to find a way of financing these activities – which led him into dealings with the early stirrings of venture capitalism, an experience at least as traumatic as anything encountered by Silicon Valley hopefuls five centuries later because it left him without ownership of the thing that he had created.

But he persevered and by about 1450 had a workable system

going. What he then needed was a product – something that would demonstrate the efficacy of his new printing technology, but for which there would also be a market. Not surprisingly (for this was still the Middle Ages) he eventually hit on the idea of printing the Bible – then as now a big hit with the religious fraternity. We don't know for sure when the first copy rolled off the Gutenberg press, but we do know that at least one copy was in existence by 1454 because in March the following year a man named Enea Silvio Piccolomini – who later became Pope Pius II – reported in a letter dated 1455 that in Frankfurt, the year before, 'a marvellous man had been promoting the Bible'. Piccolomini had seen parts of it and it had such neat lettering, he claimed, 'that one could read it without glasses'. He also claimed that 'all copies' had been sold.[4]

So let's say that printing by moveable type effectively dates from 1455. Once that genie was out of the bottle, it spread with astonishing speed. In her scholarly history of the printing revolution in Europe[5], Elizabeth Eisenstein reckoned that by 1471 there were fifteen Gutenberg-type presses in operation in Europe. By 1480, that number had increased to at least eighty-seven. By 1500, it's been estimated that something like fifteen thousand new books had been printed; by 1550 more than eight million had been printed[6] and from then onwards the avalanche of print gathered speed.

And it's still going strong. In 2008, for example, 120,947 new books were published in the UK alone,[7] and although this number is down on the all-time record year of 2003, when 129,762 titles were published, it's nevertheless an astonishing statistic. In 2009 the total number of English-language books held on the Nielsen Book database stood at over eight million globally – including in-print, print-on-demand, ebooks and out-of-print titles.

Gutenberg's legacy

But, in a way, the numbers tell only a small part of the story of the impact of printing. As Myron Gilmore put it in *The World of Humanism*, 'The invention of printing brought about the most radical transformation in the conditions of intellectual life in the history of Western civilization . . . Its effects were sooner or later felt in every department of human activity.'[8]

Here are just a few of those effects:

Mass production

In effect, Gutenberg invented what Henry Ford perfected – an industrial system for stamping out perfect copies of a standardized product. Printing was the precursor of mass production and its associated notion of economies of scale. It created new trades, professions and occupations which hadn't existed in the earlier, scribal, age.

Advertising

Before printing, advertising and publicity was mostly a word-of-mouth business. But printing provided tools – handbills, posters, flyers, pamphlets – for producing and distributing publicity materials in large quantities.

Intellectual property

The idea of intellectual property was largely unknown before the age of printing, and indeed – as Adrian Johns has pointed out[9] – it took a long time for the idea to emerge that intellectual products could, in some sense, be 'owned'. In a scribal age, the idea of a single 'author' didn't really exist, except in a metaphorical sense. (Who, for example, was the Isaiah who 'wrote' the book in the Old Testament that bears his name?)[10]

But with printing the notion of an individual author emerged in the end, and the system of 'copyright' evolved as a way of defining and protecting the contents of printed books. It was later expanded to include other forms of expression (and is now looking distinctly dysfunctional in a digital age – as we will see in Chapter 8). But it originated with printing, which then also led to the extension of the notion of intellectual property to inventions and innovations. Human beings have been inventing and innovating since the dawn of time, but before printing there was no reliable way of recognizing (and claiming ownership of) particular innovations. It was impossible to ascertain precisely what was known. 'Steady advance,' wrote George Sarton, 'implies exact determination of every previous step'[11]. Printing made this 'incomparably easier', and thus paved the way for the vast apparatus of patenting and trademarking that characterizes the modern world.

Accessibility

In the pre-Gutenberg age, books were copied by hand – generally by scribes working in scriptoria. They were thus rare and fabulously expensive objects. In 1424, for example, Cambridge University library owned only one hundred and twenty-two books, and it was reckoned that each of them had a value equal to that of a farm or vineyard. In cathedral libraries, the books were chained to prevent (or at least discourage) theft. Books were therefore the exclusive preserve of the rich and powerful – aristocrats, monasteries and medieval universities in the main. After Gutenberg, books became much more affordable and therefore plentiful (remember that estimate of fifteen thousand new titles by 1500). They became something that modestly prosperous households could afford. And this, in turn, had other powerful effects.

The Reformation

Think, for example, of the devastating impact that printing had on the Catholic Church. Before Gutenberg, the 'ordinary' people of European countries were dependent on the Church for access to, and interpretation of, the Bible – which was widely believed to be the word of God and available only in Latin, the *lingua franca* of the Church. After Gutenberg, printed Bibles were produced in increasing numbers and – perhaps more importantly – translated into vernacular languages. Suddenly, ordinary people were able to read the word of God for themselves, after which it was only a matter of time before they began to make their own interpretations of it. And then there was the astonishing way in which printing amplified the theological revolt of an obscure monk, Martin Luther, after he had pinned his Ninety-Five Theses to the door of a church in Wittenberg in October 1517.

One historian reckons that between 1517 and 1520 Luther's thirty publications probably sold well over three hundred-thousand copies. 'Lutherism,' wrote A.G. Dickens,

> was from the first the child of the printed book, and through this vehicle Luther was able to make exact, standardized and ineradicable impressions on the mind of Europe. For the first time in human history a great reading public judged the validity of revolutionary ideas through a mass-medium which used the

> vernacular language together with the arts of the journalist and the cartoonist.[12]

Protestantism represented a subtle shift of spiritual responsibility from the Church to the individual. Needless to say, the Catholic Church detested the shift. 'You should nott be your owne masters,' ranted Reginald Pole, a Catholic cardinal during Mary Tudor's reign, who held that 'household religion was a seed-bed of subversion'. Which indeed it was. At the Council of Trent in the middle of the sixteenth century the Church attempted to head off the threat posed by Gutenberg by, for example, insisting on approving 'authorized' versions of the Bible (the *imprimatur* system) and creating an 'index' of proscribed books – i.e. texts that Catholics were forbidden to read on pain of excommunication.

But at the same time the Church also tried to use printing as a tool for enforcing standardization across its vast spiritual empire. Prior to printing, liturgical texts were produced in manuscript and were therefore of variable consistency and accuracy, so local variations had to be tolerated by the authorities in Rome. But printing made it possible for the hierarchy to insist on uniform texts and rubrics so that in all European countries liturgical practice became standardized. The flip side of that coin, of course, was that local innovation and ingenuity was quashed, and so the Catholic liturgy effectively fossilized into a shape from which it was only rescued by the Second Vatican Council in the early 1960s.[13]

Scholarship

Intellectual life was transformed by printing. Before Gutenberg, if a scholar wished to consult different texts then he had no choice except to travel. After Gutenberg the same scholar could access a rapidly expanding number of texts without venturing far from his home institution. And this in turn led to massive and unexpected changes. In the scribal age, one of the major problems scholars faced was the fact that errors were an inescapable by-product of manual copying, so copies of copies became progressively degraded. After the arrival of printing, scholarship morphed from obsession with, and annotation of, single texts into an activity that was much more concerned with cross-referencing and comparison. Printing provided two things that were impossible in a scribal age: typographical fixity of texts, and the consequent possibility of cumulative development. Before Gutenberg it was difficult to build on the work of others and add to the corpus of what was already known. After Gutenberg, this kind of intellectual accumulation became the defining characteristic of scholarship.

Science

Science as we know it today is an elaborate social system involving experimentation, intense collaborative activity and incessant publication in journals, books and other outlets. Nothing like it existed in Gutenberg's time for the simple reason that it couldn't: the typographical and publishing infrastructure was lacking. Furthermore, the growth of scientific knowledge is a cumulative process in which successive generations of practitioners build on foundations established by their predecessors (with occasional outbreaks of the disorder that Thomas Kuhn called 'revolutions'[14]). None of this was possible in a pre-Gutenberg age, and so printing's provision of typographical fixity and its facilitation of cumulative development proved critical for the evolution of what became modern science. This process of standardization resulted, for example, in the publication of tables of numerical data, easily reproducible maps and charts, and in their dissemination on a Europe-wide basis. So it's no accident that many of those whom we regard as the founders of modern science – for example, Copernicus, Andreas Vesalius, Tycho Brahe, Francis Bacon, Galileo, Johannes Kepler, William Harvey and Descartes – all lived and worked during the first century of the typographical era. And printing also led to the popularization of scientific ideas because it made such ideas available to a wide public through translations into vernacular languages. The result was that by the end of the sixteenth century, the principles of geometry, anatomy, astronomy and physics were accessible to anyone in Europe who could read. The age of the alchemists was over.

Childhood

Gutenberg's invention changed our concept of childhood, as the cultural critic Neil Postman has pointed out. He argues that childhood is a protected period in the life of a person before they are seen as capable of looking after themselves. In a pre-print age, adulthood began at the point where a young person had attained communicative competence in the predominating information environment. In a society based on non-written, oral communication, that point was reached at the age of seven. As it happens, this was defined by the medieval Church as the 'age of reason' – the age at which an individual was deemed capable of being responsible for his or her actions. It also explains – says Postman – why one never sees children in Brueghel paintings: we see only small persons dressed in adult garb.

Postman argues that, in a print-based culture, it took longer for a child to attain full communicative competence: a considerable amount of schooling was required, and this pushed back the transition into full adulthood to the age of, say, twelve. The new definition of adulthood was based on *reading competence.*[15] So in this sense, the transformation in our communications environment brought about by Gutenberg even led to the extension of 'childhood'.

These are all crude summaries of examples of what one might call the 'macro' impact of Gutenberg's technology. Scholars and cultural historians would regard such summaries as perfunctory at best. But sometimes abbreviation is a product of necessity: we have a lot of ground to cover, and printing is just one part of my story. There is, however, one important thought to be borne in mind: none of these macro impacts happened overnight. I'm not trying to argue that one day printing technology arrived and the next day science (or scholarship or religious consciousness) changed; indeed, that's exactly what I'm trying to avoid. The point is that large cultural and social changes happen over extended periods of time – sometimes over centuries – and that often they are clearly visible only in retrospect. The great thing about Gutenberg's invention is that we can view its results with the 20-20 vision of hindsight.

Writers, readers and changing minds

In addition to its macro impact, printing also affected human consciousness in a more intimate way: it enabled the emergence of two individuals hitherto almost unknown – the writer and the reader.

The scribal age had little real use for the idea of the individual creative author. Postman illustrates the point by quoting Saint Bonaventura, who tells us that in the fourteenth century there were four ways of making books:

> A man might write the works of others, adding and changing nothing, in which case he is simply called a 'scribe' . . . Another writes the works of others with additions which are not his own: and he is called a 'compiler'. . . Another writes both others' work and his own, but with others' work in principal place, adding his own for purposes of explanation; and he is called a 'commentator' . . . Another writes both his own work and others', but with his own in principal place adding others' for purposes of confirmation, and such a man should be called an 'author'.[16]

Personalized authorship therefore had no place in the scribal world. There was, as Postman observes, no literary tradition of intimate disclosure, no established 'voice' or tone by which private thoughts were expressed publicly. 'Certainly there were no rhetorical conventions for addressing a throng that did not exist except in the imagination'.[17]

Someone like the great French essayist Montaigne (who was born in 1533) would therefore have come as a profound shock to St Bonaventura. He more or less invented the personal essay – short pieces which celebrated and chronicled his idiosyncratic, narcissistic, prejudiced self[18] – in an embodiment of the genre perceptively labelled four hundred years later by Norman Mailer as *Advertisements for Myself*.[19] Montaigne, says Andrew Sullivan, a leading online commentator, was living his scepticism, daring to show how a writer evolves, changes his mind, learns new things, shifts perspectives, grows older – and that this, far from being something that needs to be hidden behind a veneer of unchanging authority, can become a virtue, a new way of looking at the pretensions of authorship and text and truth.[20]

What this means, of course, is that printing and individualism are inextricably bound up with one another. It would be absurd to argue that 'individualism' was created by the printing press. But it is true that printing helped to create a social environment within which the idea of individuality made sense. And, in a way, that may have been its most profound impact on our world.

'We shape our tools,' observed Marshall McLuhan, 'and afterwards our tools shape us'. As the printed book became the dominant container for knowledge, it made special demands on those who wished to access that knowledge. Prior to printing, all human communication occurred in a social context. But with the printed book,' writes Postman,

> another tradition began: the isolated reader and his private eye. Orality became muted, and the reader and his response became separated from a social context. The reader retired within his own mind, and from the sixteenth century to the present what most readers have required of others is their absence, or, if not that, their silence. In reading, both the writer and the reader enter into a conspiracy of sorts against social presence and consciousness.[21]

In that sense, reading is an anti-social act which imposes distinctive demands on the reader: immobility; isolation; silence; concentration; the ability to *immerse* oneself in the thought processes of the writer and to remember and make links with the thoughts of other writers as expressed in other texts; and so on. These are not abilities that are genetically determined: they have to be taught – and learned. This is why Postman argues that printing changed our conception of education, extending it beyond the 'Age of Reason' (seven years as defined by the medieval Church) to twelve or even fourteen – and thus to the extension of 'childhood'.[22]

The fact that we are not genetically programmed for reading in the way that we are for language is very significant because it may have shaped the structure of our brains. In her path-breaking exploration of the neuroscience of reading, Maryanne Wolf points out that human beings invented reading only a few thousand years ago and that this invention actually changed the way our brains are organized, which in turn altered the way our species evolved.[23]

Wolf's perspective is informed by what neuroscientists have discovered in recent decades about the astonishing plasticity of the brain. Every time we acquire a new skill, groups of neurons in the brain create new connections and pathways among themselves. So our brains have what computer scientists might call an *open architecture* – one that is versatile enough to reconfigure itself in response to changing circumstances. This implies that the 'reading brain' is the product of a two-way process:

> Reading can be learned only because of the brain's plastic design, and when reading takes place, that individual brain is forever changed, both physiologically and intellectually. For example, at the neuronal level, a person who learns to read in Chinese uses a very particular set of neuronal connections that differ in significant ways from the pathways used in reading English.

> When Chinese readers first try to read in English, their brains attempt to use Chinese-based neuronal pathways. The act of learning to read Chinese characters has literally shaped the Chinese reading brain.[24]

Wolf's account of how reading changes our brains lends a special piquancy to McLuhan's phrase: *and afterwards our tools shape us*. But it's just the latest milestone in an inquiry that has occupied scholars over four centuries, from Erasmus to Eisenstein. The consensus they have reached, according to Postman, is that reading fosters rationality, that the form of the printed book encourages what Walter Ong called 'the analytic management of knowledge'.[25] Engaging with a printed text, he says, requires considerable powers of classifying, inference-making and reasoning:

> It means to uncover lies, confusions and over-generalizations, to detect abuses of logic and common sense. It also means to weight ideas, to compare and contrast assertions, to connect one generalization to another. To accomplish this, one must achieve a certain distance from the words themselves, which is, in fact, encouraged by the isolated and impersonal text.[26]

Now of course there's an element of nostalgic hyperbole here: analytic thinking existed long before Gutenberg, and print-based culture also fostered a good deal of frivolity, obscurantism and worse. But there's a kernel of truth in Postman's summary of the scholarly consensus, and we will see it more clearly in a moment when we consider how our – post-Gutenberg – contemporary communications technology might be shaping us. What Postman *et al* are saying, in effect, is that printing spurred the evolution of what might be called *Homo Typographicus*. Which in turn raises the question: are we now embarking on the evolution of *Homo Interneticus*? And if so, what might this creature be like?

After Gutenberg, what next?

Printing provides the best-documented case study we have of what happens when our media environment undergoes a revolutionary change. What it shows is that the technology had a profound impact on human society, and that its effects went way beyond anything that could have been foreseen in the early years of the revolution.

In a way, we ought not to be terribly surprised by this. The word 'media' is the plural of *medium*, a word with an interesting etymology. The conventional – journalistic – interpretation holds that a medium is a carrier of something. But in science, the word has another, more interesting, connotation. To a biologist, for example, a medium is a mixture of nutrients needed for cell growth. And that's a very interesting interpretation for our purposes.

In biology, media are used to grow tissue cultures – living organisms. The most famous example, I guess, is Alexander Fleming and the mould growing in his Petri dishes which eventually led to the discovery of penicillin.

It seems to me that this is a useful metaphor for thinking about human society; it portrays our social system as a living organism which depends on a media environment for the nutrients it needs to survive and develop. Any change in the environment – in the media which support social and cultural life

– will have corresponding effects on the organism. Some things will wither; others may grow; new, unexpected species may appear. The key point of the metaphor is simple: change the environment, and you change the organism; change the media environment and you change society.

Printing changed our environment – and it shaped human society for the next four hundred years. If the Internet is as revolutionary a technology as printing was, then we can expect impacts on a comparable scale, and we have to accept that we are currently not in a position to judge what its long-term effects will be. But that doesn't stop us speculating.

Utopianism

At the *macro* level, ever since the Internet appeared on the radar of public consciousness it has been the focus of Utopian dreams and dystopian nightmares about how it might change society. The dreams stemmed from the realization that the network was, somehow, more than just a network: that it was a kind of virtual *place*. In due course, this acquired a name – *cyberspace* – a term first used in a vague sense by the writer William Gibson in his novel *Neuromancer*[27] and eventually given an effective definition by Bruce Sterling in his book *The Hacker Crackdown*:

> Cyberspace is the 'place' where a telephone conversation appears to occur. Not inside your actual phone, the plastic device on your desk. Not inside the other person's phone, in some other city. The place between the phones. [. . .] in the past twenty years, this electrical 'space', which was once thin and dark and one-dimensional – little more than a narrow speaking-tube,

> stretching from phone to phone – has flung itself open like a gigantic jack-in-the-box. Light has flooded upon it, the eerie light of the glowing computer screen. This dark electric netherworld has become a vast flowering electronic landscape. Since the 1960s, the world of the telephone has cross-bred itself with computers and television, and though there is still no substance to cyberspace, nothing you can handle, it has a strange kind of physicality now. It makes good sense today to talk of cyberspace as a place all its own.[28]

In the early days of the Internet – say between 1983 and 1993 – this metaphor of cyberspace exerted a powerful hold on those who used the network. It gave them a powerful vision of 'a new frontier, where people lived in peace, under their own rules, liberated from the constraints of an oppressive society and free from government meddling'.[29] This vision was reinforced by the fact that the early Internet was a space in which corporations and commercial forces were largely absent.

So it's perhaps not surprising that the intoxicating newness of the virtual space fuelled an upsurge of libertarian Utopianism. This was a time when people believed that the network might really challenge the authority of nation-states. As Goldsmith and Wu put it, the notion of cyberspace,

> tapped into something much deeper, causing many to hope that the new network might really change things, somehow liberate us from the world we live in and even do something to change the human condition. Behind every vision of Internet Utopianism lay the hope that connecting every human being on earth might make the world a better place. Humanity united might do better than our lousy systems of government, throw away the concept

of the nation-state, and live in some different but better way.[30]

The apogee of this wave of libertarian Utopianism was John Perry Barlow's celebrated *Declaration of the Independence of Cyberspace*.[31] It was prompted by the Communications Decency Act of 1996, the first attempt by US lawmakers to control Internet communications.[32] Taking his cue from the sonorous tone of the original US Declaration of Independence, Barlow wrote:

> Governments of the Industrial World, you weary giants of flesh and steel, I come from Cyberspace, the new home of Mind. On behalf of the future, I ask you of the past to leave us alone. You are not welcome among us. You have no sovereignty where we gather.
>
> We have no elected government, nor are we likely to have one, so I address you with no greater authority than that with which liberty itself always speaks. I declare the global social space we are building to be naturally independent of the tyrannies you seek to impose on us. You have no moral right to rule us nor do you possess any methods of enforcement we have true reason to fear.
>
> Governments derive their just powers from the consent of the governed. You have neither solicited nor received ours. We did not invite you. You do not know us, nor do you know our world. Cyberspace does not lie within your borders. Do not think that you can build it, as though it were a public construction project. You cannot. It is an act of nature and it grows itself through our collective actions.
>
> You have not engaged in our great and gathering conversation, nor did you create the wealth of our marketplaces. You do not know our culture, our ethics, or the unwritten codes that already provide our society more order than could be

> obtained by any of your impositions.
>
> You claim there are problems among us that you need to solve. You use this claim as an excuse to invade our precincts. Many of these problems don't exist. Where there are real conflicts, where there are wrongs, we will identify them and address them by our means. We are forming our own Social Contract. This governance will arise according to the conditions of our world, not yours. Our world is different.

With the 20-20 vision of hindsight, it's easy to patronize the idealism (or naiveté?) implicit in Barlow's ringing declaration. Subsequent decades have shown that the Internet does not always lie beyond the reach of the nation-state, that the threat of its domination by large companies like Google, Apple, Amazon and Microsoft has grown and that cyberspace has become a battleground for powerful industrial interests. But the Utopianism of the founders and early users endures as a stubborn undercurrent which surfaces at critical moments when online freedoms or network 'neutrality' seem threatened – as for example in late 2010 when anonymous groups of hackers launched retaliatory cyber-attacks against companies that terminated services to WikiLeaks. And an increasing number of people have bought into this – to the extent that the 'freedom to connect' is coming to be seen as a basic human right that the 'weary giants of flesh and steel' should not attempt to deny.[33]

Utopianism lives on in other, more pragmatic, manifestations too – for example in the idea of the synergistic benefits that spring from enabling countless intelligences to join forces in ways that demonstrate the 'wisdom of crowds'[34]; or to collaborate in vast networked enterprises like **Open Source** software[35] and Wikipedia; or to create and publish cultural artifacts (photographs, blogs, home-made movies, songs) on a hitherto unimaginable scale.[36]

Dystopianism

The Communications Decency Act of 1996 may have been the first legislative response to the Net, but it had been the subject of dystopian nightmares even before that curious statute emerged from beneath a Congressional stone. And in the decade and a half since the Supreme Court struck down the part of the act relating to 'indecency', fears about the network's social impact have proliferated in sync with the surprises it has sprung on an unsuspecting world.

These fears are so widespread and diverse that they almost defy summarizing,[37] but the main themes include: a conviction that the network is reshaping our intellectual, social, economic and political landscape in unpalatable ways; a belief that ubiquitous networking is changing our conceptions of art and entertainment – and blurring the distinction between news and entertainment; a perception that the Internet is fragmenting our culture into bite-sized chunks, overwhelming us with data, eroding personal privacy, polarizing our politics. The network, we are told, is creating a world of atomized, isolated individuals who would sooner send an email to a colleague in the next-door cubicle than lean over to talk to her. It has unleashed upon us an avalanche of pornography, misinformation, lies and idle chatter. It has become a platform for wholesale piracy of intellectual property and a destroyer of venerable business models which have served society and democracy well. It has become a godsend for Islamic terrorists and extremists of every stripe, and a patchwork of Balkanized 'echo-chambers' in which people only hear the opinions of those with whom they agree.

And, of course, it has – allegedly – destroyed the art of reading.

The human impact

In addition to these conjectures about the network's 'macro' impact, we're seeing the rise of speculation about how a comprehensively networked media environment may be affecting the ways we think, and indeed even how our brains work.

The most evocative articulation of this came in an article by the technology commentator Nicholas Carr in the July/August 2008 issue of the *Atlantic*.[38] Under the headline 'Is Google Making Us Stupid?' Carr confessed that over the past few years he's had,

> an uncomfortable sense that someone, or something, has been tinkering with my brain, remapping the neural circuitry, reprogramming the memory. My mind isn't going – so far as I can tell – but it's changing. I'm not thinking the way I used to think. I can feel it most strongly when I'm reading. Immersing myself in a book or a lengthy article used to be easy. My mind would get caught up in the narrative or the turns of the argument, and I'd spend hours strolling through long stretches of prose. That's rarely the case anymore. Now my concentration often starts to drift after two or three pages. I get fidgety, lose the thread, begin looking for something else to do. I feel as if I'm always dragging my wayward brain back to the text. The deep reading that used to come naturally has become a struggle.

As it happens, the striking title of Carr's essay is misleading, because his main target is not the world's leading search engine but our networked media ecosystem. What the Net seems to be doing, he says, is 'chipping away my capacity for concentration and contemplation'. His mind now expects to take in information 'in a swiftly moving stream of particles'. Once he was 'a scuba diver in the sea of words'. Now he whizzes along the surface 'like a guy on a Jet Ski'.

Carr maintains that his experience is characteristic of that of many other people. He cites, for example, Bruce Friedman, a pathologist at the University of Michigan Medical School, who blogs regularly about the use of computers in medicine, and who has written about how the Internet has altered his mental habits:

> I now have almost totally lost the ability to read and absorb a longish article on the web or in print. To create my blog notes, I browse various web resources and write relatively short notes such as this one. I also restrict my blog notes to about 200–300 words, assuming that my readers have a short attention span similar to mine.[39]

Friedman later told Carr that his thinking has taken on a 'staccato' quality, reflecting the way he quickly scans short passages of text from many sources online. 'I can't read *War and Peace* anymore. I've lost the ability to do that. Even a blog post of more than three or four paragraphs is too much to absorb. I skim it.'[40]

Widespread public interest in Carr's article encouraged him to expand the thesis into a full-length book.[41] Not everyone has been impressed by his arguments, though. The Pew Internet and American Life project, for example, asked the panel of over 370 experts who have participated in its annual 'Future of the Internet' surveys for their reaction to Carr's hypothesis. Eighty-one per cent of them disagreed with Carr's thesis.

The public debate sparked by Carr's essay has been lively and thoughtful,[42] which suggests that he may indeed have echoed some deep-seated anxieties. Most of his critics concede that he has hit on an important subject, and agree that it is likely that the Internet will have far-reaching effects on our culture. But beyond that, the consensus evaporates. The writer and futurologist Jamais Cascio, for example, argues that human cognitive capacities have always evolved to meet new challenges in our environment. One of our contemporary challenges, he points out, is the torrent of information unleashed by the Net. But this proliferation of diverse voices may have the effect of improving our ability to think.

In *Everything Bad Is Good for You*, Steven Johnson argues that the increasing complexity and range of media we engage with has, over the past century, made us smarter, rather than dumber, by providing a form of cognitive calisthenics. Even pulp-television shows and video games have become extraordinarily dense with detail, filled with subtle references to broader subjects, and more open to interactive engagement. They reward the capacity to make connections and to see patterns – precisely the kinds of skills we need for managing an information glut.[43]

Cascio describes these skills as 'fluid intelligence' – the ability to find meaning in confusion and to solve new problems, independent of acquired knowledge. This is different from the abilities that people have hitherto associated with intelligence –

the capacity to memorize and recite factual information. But building it up may improve the capacity to think deeply that Carr and others fear we're losing for good. And we shouldn't let the stresses associated with a transition to a new era blind us to that era's astonishing potential. We swim in an ocean of data, accessible from nearly anywhere, generated by billions of devices. We're only beginning to explore what we can do with this knowledge-at-a-touch.[44]

In this view, the scattergun mentality decried by Carr may turn out to be a short-term problem. The problem isn't that we have too much information at our fingertips, Cascio thinks, but that our tools for managing it are still in their infancy. He points out that worries about "information overload" predate the rise of the Web and that many of the technologies that Carr obsesses about emerged to help humanity cope with a flood of ideas and data. In that sense, Google may not be the problem but the beginning of a solution.

This view – that, far from making us stupid, networking technology will actually wind up making us smarter, that the Internet is (so to speak) 'power-steering for the mind', or a technology for realizing Doug Engelbart's dream of using computers to 'augment' human intelligence[45] – is an enduring theme of those who are critical of Carr's dystopian concerns. Underlying these more favourable views of networking technology is an assumption that humanity is trapped in a cybernetic spiral in which technological development leads to more complex information environments, which in turn require more powerful tools to enable humans to deal with the complexity, which in turn make even more demands on us, which in turn requires . . . Well, you get the idea.

* * *

While this debate about the macro impact of the Internet rages, people are also beginning to speculate about how our plastic, adaptive brains are being reshaped by ubiquitous networking. In a way, this is just the contemporary echo of Maryanne Wolf's study of how the structure of the human brain was shaped by print. As with Wolf, the emphasis is on how the habits inculcated – or encouraged – by online activity may affect cognitive processes. Reading online is not just about reading text in isolation, writes Gloria Mark, a human-computer interaction (HCI) specialist at the University of California, Irvine:

> When you read news, or blogs or fiction, you are reading one document in a networked maze of an unfathomable amount of information. My own research shows that people are continually distracted when working with digital information. They switch simple activities an average of every three minutes (e.g. reading email or IM) and switch projects about every 10 and a half minutes. It's just not possible to engage in deep thought about a topic when we're switching so rapidly.[46]

What this suggests to some theorists is that the endless hyperlinked diversions available on the Web are not just seductive but may even be addictive. In support they cite experimental work by neuroscientists like Jaap Panksepp of Washington State University and Kent Berridge of the University of Wisconsin.[47] Panksepp believes that humans are genetically programmed for seeking, in that when we get thrilled about the world of ideas, about making intellectual connections, about divining meaning, it is the seeking circuits in our brains that are firing.

A key factor in this process is alleged to be dopamine – an endogenous chemical in the brain which relays, amplifies, and modulates signals between a neuron and another cell. It seems that the dopamine circuits promote what the researchers call 'states of eagerness and directed purpose'. These are states that humans love to be in – so that we seek out substances (like cocaine and amphetamines) that promote them. The conjecture is that we also seek out *activities* that have a similar effect. So, writes Emily Yoffe,

> Ever find yourself sitting down at the computer just for a second to find out what other movie you saw that actress in, only to look up and realize the search has led to an hour of Googling? Thank dopamine. Our internal sense of time is believed to be controlled by the dopamine system. People with hyperactivity disorder have a shortage of dopamine in their brains, which a recent study suggests may be at the root of the problem. For them even small stretches of time seem to drag. [48]

At the moment, all of this is pure speculation based mostly on long chains of extrapolation from animal experiments. This line of inquiry may turn out in the end to be hogwash. But curiosity about the longer-term impact of the Net's transformation of our communications environment seems to me to be a sensible preoccupation. As Clay Shirky puts it: 'It is too early to tell whether the Internet's effect on media will be as radical as that of the printing press. It is not too early to tell that there is nothing that happened between 1450 and now that comes close.'[49]

Which is why it's worth trying to learn from Gutenberg. Mankind's experience with print technology has taught us a few lessons that may be useful in contemplating our networked future. It shows, for example, how a revolutionary change in communications technology can, in the long run, radically transform the society in which it occurs. It confirms the view that while the short-term impact of the new technology may be fairly easy to predict, its longer-term impacts are not. But perhaps the most intriguing lesson of the Gutenberg experiment is the thought that, in addition to reshaping society, a dominant communications technology may also reshape *us*.

2. THE WEB IS NOT THE NET

OK, here's an outrageous assertion: *we still don't understand the Net*. As you can imagine, when I say this in lectures it sometimes causes audiences to bridle with indignation. After all, my listeners are intelligent and knowledgeable people, many of whom would describe themselves as 'Internet-savvy'. So how dare I make the claim that they don't understand what I'm talking about!

I'm serious, though. There are two ways in which it's clear that our society still hasn't understood the Internet. The first comes under the heading of glaring misconceptions – for example the delusion (common in mainstream media folks in the early days) that it's just another **broadcasting** medium. That was nonsensical, for reasons that we will come to in Chapter 4. The fact that serious media executives once believed it we can put down to a combination of two parts ignorance and one part wishful thinking.

But the most common – and still widespread – misconception is the belief that the Internet and the Web are synonymous.[1] So let's immediately feed that particular red herring to the nearest available cat. A good way to think about it is to use a railway metaphor. Think of the Internet as the tracks and signalling technology of the system – the infrastructure on which everything runs. In a railway system different kinds of traffic run on the infrastructure: high-speed express trains, slow stopping trains, commuter trains, freight trains and (sometimes) specialist maintenance and repair trains.

In the Internet context, web pages are just one of the many kinds of traffic that run on the infrastructure. Other kinds of traffic include: music files being exchanged via **peer-to-peer** networking; software updates; email; instant messages; phone conversations via Skype and other **VoIP** (voice over **Internet Protocol**) services; **streaming media** (video and audio); and other stuff too arcane to mention. And – here's the important bit – there will be other kinds of traffic, stuff we haven't dreamed of yet, running on the Internet in ten years' time. So the thing to remember is this: the Web is huge and very, very important – *but it's just one of the things that run on the Internet*. The network is much bigger – and far more important – than anything that runs on it. Understand that and you're half-way to wisdom.

The tendency to identify a technology with the particular instantiation of it with which one happens to be familiar is an old story. In the 1930s, for example, a radio or wireless *receiver* would usually be referred to as a 'radio' or 'the wireless'. In the 1950s the first portable radios powered by *transistors* rather than by **thermionic valves** appeared and were properly described as *transistor radios*, but they became universally known as 'transistors'. Similarly in the 1970s *videotapes* became 'videos'.

But these were pretty innocuous usages because they didn't obscure anything important. If, however, people confuse the Internet with the **application** that they happen to use most intensively – whether it be Google, Facebook or the Web – then they are missing something really important about the Internet. To see what it is, go to Chapter 3.

3. FOR THE NET, DISRUPTION IS A FEATURE, NOT A BUG

I've said that my impression is that most people still don't understand the Net. Actually, there is a better concept than 'understanding' which gets closer to what I mean: it's *appreciation* – defined as 'grasping the nature or essence of something'. So my claim is really that we don't appreciate the Net. In fact our attitude to it has undergone a weird transition in the last two decades. Somehow the Internet went from being something exotic and mysterious to being something that we take for granted, like electricity or running water. I remember having conversations about the Net (and the burgeoning Web) with newspaper editors and other 'thought leaders' in the mid-1990s in which it was clear that they regarded it as something that belonged firmly in the specialized world of boffins, nerds and 'anoraks'. 'This Web thingy,' said one national newspaper editor to me in 1995, 'don't you think it's just the Citizen Band radio *de nos jours*?' In 1999, Andy Grove, then the CEO of the chip-maker Intel, was widely ridiculed for his prediction that 'companies that aren't Internet companies within five years won't be companies at all'.[1] Did he mean that every cobbler and fast-food joint in the universe would have to be online?

In fact, Grove was just being perceptive. What he meant was that the Internet would become like mains electricity or the telephone – something that all businesses would have to use and reckon with. He was speaking at a time when companies still used to issue delighted press releases announcing that they were 'on the Web' – much as companies once boasted that they were 'now on the telephone'.

And in fact the transition envisaged by the Intel CEO did indeed come to pass. What infuriates me as a professional observer of these things is that I can't now say when the tipping point was reached – I missed the moment when the popular perception of the Internet switched from exoticism to mundanity. Was it the first time a new PC came with a (Microsoft) web **browser** pre-installed? Or the first Christmas that people started buying presents on Amazon? Or the moment when the only way you could book a flight on a budget airline was to go online? Whenever the switch occurred, however, one thing has remained constant, namely our lack of curiosity about the network. Once, it was too weird to be worth trying to comprehend; now, it's too boring to be bothered with. The net result is that we have been sleepwalking into the future. We have come to depend on a global communication system about which most of us have little knowledge and even less curiosity. And that's bad news, for all kinds of reasons, one of which is that it might lead us to allow clueless or malevolent legislators to screw it up, and another is that it might lead us, as consumers, to destroy it. But more of that later.

So what's special about the Internet?

If you're a scholar who specializes in these matters, the answer to this question is deceptively simple: the Internet is special because it's a powerful enabler of disruptive innovation – defined as 'a process by which a product or service takes root initially in simple applications at the bottom of a market and then relentlessly moves 'up market', eventually displacing established competitors'.[2] Translation: the network is a global machine for springing surprises – good, bad and indifferent – on us. What's more, it was explicitly designed to be like this, though its designers might not have expressed it in precisely those terms. In other words, the disruptiveness of the Internet is a *feature*, not a bug – it stems from the basic architectural principles of the network's design. It's what the network was designed to do. So you could say that disruptiveness is built into its virtual DNA.

To understand this, it's helpful to know something of the history and evolution of the system – and if you're really interested you'll find a detailed account in the Appendix at the end of the book. But if you're busy then all you need to know is that the uniqueness of the Internet stems from two fundamental principles that underpinned its design. These principles look innocuous, but they have had the most far-reaching implications, as we shall see in a moment.

The Internet that we use today was switched on in January 1983. It was designed over the course of the previous decade, and was derived from an earlier network, the ARPAnet, that was created by the US Department of Defense in the period 1966–72.

The 'internetworking' project that began in 1973 was led by two engineers, Vinton (Vint) Cerf and Robert Kahn, both of whom had worked on the ARPAnet. The design problem they faced was simple but profound: how do you create a network that will seamlessly link other networks and be, in some sense, future-proof? Or, more precisely, how do you design a system for applications that nobody has yet dreamed of?

The answer that Cerf and Kahn came up with was brilliantly simple. It had two components:

- There must be no central control.
- The network should not be optimized for any particular application.

In a way, both principles were based on the lessons of history, and in particular on experience with telephone networks.

No central control

As far as *control* was concerned, until the 1980s virtually all telephone networks were owned and tightly regulated either by governments or by state-regulated telephone companies which were able to dictate (a) who could – and could not – connect to the network, and (b) what the network could be used for. In the US, the state-regulated monopoly telephone provider was AT&T. The powers of such monopolies led to all kinds of absurdities – like the 'Hush-a-Phone' example discussed by Lawrence Lessig in his book *The Future of Ideas*. The Hush-a-Phone was a plastic device that could be clipped onto a telephone handset to reduce extraneous noise in, say, a crowded room. It had no electronic components. In effect, it mimicked what most of us do when we cup a hand round a handset in order to reduce the amount of ambient noise that reaches the microphone.

When the Hush-a-Phone was released on the market, AT&T

objected on the grounds that it was a 'foreign attachment.' The regulations under which AT&T operated forbade any foreign attachments without the company's permission. It had not given Hush-a-Phone any such permission. The regulator (the Federal Communications Commission) agreed with AT&T. End of Hush-a-Phone.

This was, Lessig writes,

> an extreme case. The real purpose of the foreign attachments rule was, at least as AT&T saw it, to protect the system from dirty technology. A bad telephone or a misbehaving computer attached to the telephone system could, AT&T warned, bring down the system for the whole region. Telephones were lifelines, and they had to be protected from the experiments of an inquisitive nation. Rules such as the foreign attachments rules were intended to achieve this protection.
>
> Whatever their intent, however, these rules had an effect on innovation in telecommunications. Their effect was to channel innovation through Bell Labs [the research and development arm of AT&T]. Progress would be as Bell Labs determined it. Experiments would be pursued as Bell Labs thought best. Thus telecommunications would evolve as Bell Labs thought best.[3]

The history of technology is littered with cautionary tales about what happens when powerful companies are allowed to control the pace of innovation. In his history of the communications industries in the twentieth century, Tim Wu tells an even more scarifying story about how AT&T suppressed a really significant innovation because it did not suit its own narrow commercial interests.

In this case, the technology was that of magnetic recording tape – a technology that would eventually transform the music industry and provide the basis for the mass storage devices (tape and disk drives) of the entire computer industry. Wu reveals that by early 1934 Clarence Hickman, an engineer in Bell Labs had developed a telephone answering machine. If a phone call went unanswered, Hickman's machine would beep and switch on a recording device that would enable the caller to leave a message. This was revolutionary enough for its time, but it wasn't the most significant aspect of the invention. That was the way Hickman's machine recorded the audio message – onto magnetized tape. And yet this astonishing innovation only came to light again in the 1990s.

How come? As Wu tells it, soon after Hickman had demonstrated his invention, AT&T ordered the labs to cease all research into magnetic storage, and Hickman's research was suppressed and concealed for more than sixty years, and only came to light when the historian Mark Clark came across his lab notebook in the Bell archives.[4]

Why did AT&T want to bury such an important and commercially valuable invention? The answer unearthed by Clark from company memoranda was startling: AT&T management believed that the answering machine, with its magnetic tapes, would lead the public to abandon the telephone.

The reasoning behind this was truly surreal. Bell executives believed that the knowledge that it was possible to record a telephone conversation would have catastrophic consequences for its core business. Businessmen, for example, might fear the potential use of a recording to undo or undermine a written contract. Tape recording would also inhibit use of the telephone for obscene or ethically dubious conversations – which

AT&T executives had reason to suspect constituted a significant proportion of telephone conversations, even in the 1930s. So it came about, as Wu puts it, that 'the enlightened monopolist can occasionally prove a delusional paranoid'. It was safer to shut down a promising line of research than to risk the Bell system.

Another cautionary tale about the dangers of letting network owners determine the pace of innovation was provided by the history of the fax machine. Facsimile transmission over electrical circuits is in fact a very old technology (the fax machine was patented in 1843), and specialized machines for transmitting 'wire photographs' were in regular use in the newspaper world from the 1920s. Yet it was not until the 1980s that fax machines became an accepted part of office life. Why?

Some of the reasons have to do with the time it took for the Japanese electronics industry to develop the manufacturing capability to mass-produce fax machines as consumer products. But another reason was that, until the 1980s, many telephone networks in industrialized countries were owned and operated by national post offices – which, understandably, did not look favourably on the idea of people being able to send letters over the telephone network.

As before, the moral of the story was simple: if you allow central control of a network, then *innovation will proceed at the speed deemed suitable by the controller*, rather than by the inventiveness of outsiders.

Neutrality towards applications

The second design principle was also partly a product of experience with the telephone network. The phone system was optimized for a particular application, namely voice calls. This determined both its technical architecture and its operation. In analogue communications, a voice call requires a physical connection between caller and respondent. This is achieved by a set of switches setting up the connection – and maintaining the circuit in existence for the duration of the call. This was fine for human communication (because people tend to talk quite a lot and the line is therefore being used fairly intensively) but hopelessly inefficient for computer-to-computer communication (where there are long periods of silence interspersed with short bursts of activity). Also, the average duration of a computer-to-computer 'conversation' was considerably greater than that of a voice call – requiring those expensive circuits to be maintained for longer. The design lesson to be drawn from this was that if a network is optimized for one application (voice), it may be sub-optimal for a new application (computer communication).

This led Cerf and Kahn and their colleagues to conclude that the Internet should be agnostic as to applications, and so they came up with a network that essentially did only one thing – it took data packets in at one end and did its best to deliver them to their destinations. But it was entirely indifferent to what was in the packets; it didn't care whether they were fragments of an email message, a digitized voice conversation, a pornographic video or a music track: if you had an idea for a new application that could be achieved using data packets, then the network would do it for you, with no questions asked. Over time various phrases emerged for describing this design principle – 'stupid network, smart applications' was one I remember – but in the end it became known as the 'end-to-end' principle.

These two axioms – absence of central control and the end-to-end principle – are the bedrock of the Internet's architecture.[5] And they had consequences beyond even the wildest dreams of the engineers who first proposed them. In essence, they enabled a remarkable explosion of innovation, as inventors and entrepreneurs thought up applications that could harness the capabilities of the network. They used it to spring surprises on us. Some of these surprises have been pleasant; some have been unpleasant. But all have had the effect of transforming our information environment and our economies.

In order to illustrate the point, I've focused on a number of these surprises. They fall into two categories: *first-order surprises*, by which I mean innovations that sprang directly from the Internet's open architecture. These are the World Wide Web, the Napster file-sharing system, and **malware** – the plague of malicious software that has engulfed us.

Then there are what I call *second-order surprises*, namely innovations which built on the openness of the first-order surprises. Two in particular stand out, both of which were built on the openness of the Web. They are Wikipedia and Facebook.

Let's look at them in turn.

First-order surprises

The Web

In September 1984 a young British computer scientist named Tim Berners-Lee arrived at CERN, the European high-energy physics laboratory in Geneva. He had been awarded a fellowship to work on 'data acquisition and control' in the lab's Central

Computing Division. He chose to focus on documentation systems – software that allows documents to be stored and later retrieved. This was a hot topic at CERN, a complex organization with a large, constantly changing cast of visiting research physicists working on elaborate experiments, all of which generated vast volumes of paper. Berners-Lee began to think about what would be involved in creating a documentation system that would be acceptable to all the diverse intellectual and national cultures that converged on the laboratory. He concluded that he would have to invent 'a system with common rules that would be acceptable to everyone. This meant as close as possible to no rules at all.'[6]

Berners-Lee spent the next two years assembling the ideas that would need to be combined to produce the sort of documentation system that he thought might be needed.

First of all, it would have to be completely decentralized. Secondly, it should use the idea of **hypertext** – i.e. documents which had internal links to other documents. This was already a well-established idea in the computing business and indeed Apple had launched a product, HyperCard, which implemented the idea on a personal computer. But the problem with the Apple product – and with all other hypertext implementations – is that all the links were between documents residing on a single computer. This led to Berners-Lee's third design requirement – that his system would link documents across the global Internet.

By the closing months of 1988, he had marshalled his ideas into a coherent plan. He spoke to his boss, Mike Sendall, about it, and was advised to formulate his thoughts as a project proposal. So he did. See the reproduction of the front cover opposite.

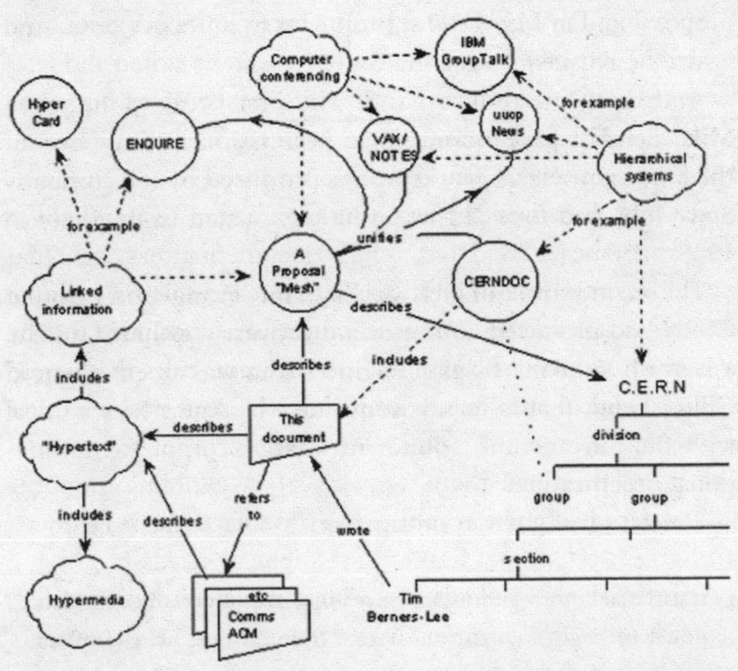

Figure 3.1: Cover page of the original Web proposal. 'Vague but exciting' was his boss's scribbled reaction.

In March 1989 he delivered the proposal to Mike Sendall, and to members of the central committee that oversaw the coordination of computers at CERN. Sendall scribbled 'vague but exciting' on his copy of the proposal. It was just about the only response Berners-Lee received. 'Nothing happened,' he wrote later.

Like a dog worrying at a bone, however, Berners-Lee kept working – entirely unofficially – on his idea. He went round CERN talking about the concept of a global hypertext system to anyone who would listen. But by early 1990 he still had received no official reaction. So he reformatted the original proposal and in May 1990 submitted it to his boss's boss. 'And again,' he recalled, 'it got shelved'.

And then fate took a hand. The first break came when Mike Sendall gave Berners-Lee permission to buy two of the fancy new NeXT workstations produced by the company Steve Jobs had founded when he was ousted from Apple in 1985.

The second stroke of luck was that Tim's evangelizing around CERN had produced an important convert – Robert Cailliau, a Flemish-speaking Belgian engineer who was an experienced CERN hand, that is to say, someone who knew how to deal with the bureaucratic politics of a large, complex, multinational organization.

'Robert's real gift was enthusiasm,' wrote Berners-Lee,

> translated into a genius for spreading the gospel. While I sat down to begin to write the Web's code, Robert, whose office was a several-minute walk away, put his energy into making the WWW project happen at CERN. He rewrote a new proposal in terms that he felt would have more effect. A CERN veteran since 1973, he lobbied among his wide network of friends throughout the organization. He looked for student helpers, money, machines and office space.[7]

To create the Web, Berners-Lee had to innovate on four separate but related fronts. Specifically, he had to:

- Invent a way of giving every single web page a unique identifier and a machine-readable address. In his original design, Berners-Lee proposed the Universal Resource Identifier (**URI**) as the identifier and Uniform Resource Locator (**URL**) as the address. For a long time they were widely regarded as synonymous, but they're not.[8]

- Design a technical **protocol**, i.e. a set of computer-readable conventions that would enable web clients and servers to communicate without ambiguity as they requested and served documents. This emerged as the Hypertext Transport Protocol (**HTTP**) – which is why the full address of a web page is prefaced by the string 'http://'.
- Create software that would enable people to browse and edit web pages (in computerspeak the '**client**' software) and a '**server**' program that would enable a networked computer to serve up web pages on demand. Browsers like Firefox, Safari, Internet Explorer and Opera are examples of client software.
- Come up with a standard language for marking-up web pages so that they would be rendered consistently by any browser. For this purpose Berners-Lee adapted SGML (Standard Generalized Markup Language), a widely used standard in the publishing industry. His variant he called **HTML** – for HyperText Markup Language.

What's interesting about this list is that, in normal circumstances, it would provide a reasonable work-plan for a group of four talented software engineers working intensively over a period of at least a year. The truly astonishing thing is that, working almost entirely on his own,[9] Berners-Lee came up with a working system in a few months. He began writing the code for the Web in October 1990. By mid-November he had a browser/editor and the first version of a web server running on his NeXT workstation. By Christmas Day 1990, the browser/editor was working on Berners-Lee's and Cailliau's

NeXT machines, communicating over the Internet with the world's first web server, info.cern.ch.[10]

The next few months were spent evangelizing the Web to colleagues within CERN. A shrewd tactical move was to put the one reference document that everybody used every day – the lab's telephone directory – on the nascent Web. Nicola Pellew's simple **browser** was 'ported' (computerspeak for 'translated to run on') to all sorts of other machines, from mainframes through **Unix** workstations to plain DOS for the [IBM] PC. 'These were not great showcases for what the Web should look like,' Berners-Lee wrote, 'but we established that no matter what machine someone was on, he would have access to the Web.'[11]

In March 1991, the Web was extended to the limited number of people in CERN who had NeXT workstations. In May the physicist Paul Kunz from the Stanford Linear Accelerator (SLAC) arrived at CERN and was shown the Web. Since he was a keen user of the NeXT workstation he was easy to convince, and when he returned to Stanford in the summer the first web server outside of Europe was set up at SLAC in Palo Alto, California.

The extraordinary thing is that while all of this was going on, the Web was never an official CERN project. Repeated bids to the lab's management for resources – people, computers, office space – were ignored or turned down. Although this may seem crazy in retrospect, in fact it was perfectly understandable. CERN's business was particle physics, not computer science. It was a hugely expensive facility, and the national governments who funded it might not have taken kindly to the idea that some of their money was being used to support some wacky computing project.

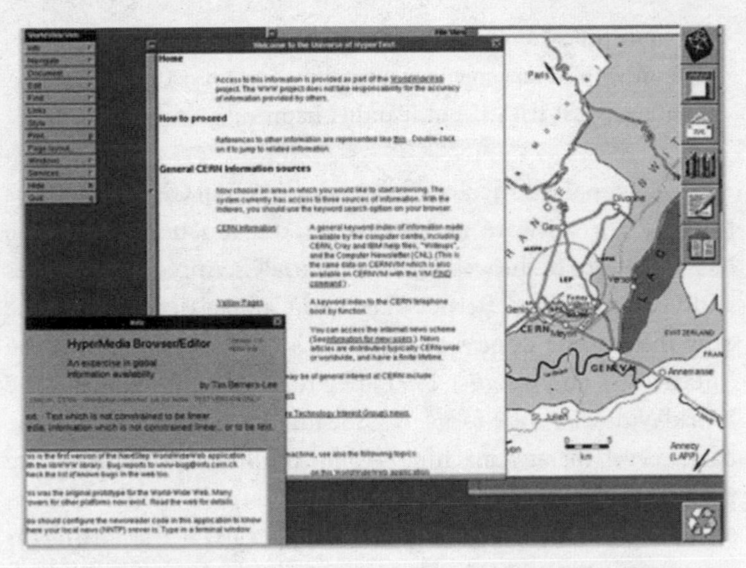

Figure 3.2: Screenshot of an early version of the browser/editor client program running on Tim Berners-Lee's NeXT workstation. (Source: CERN http://info.cern.ch/NextBrowser.html)

But at the same time, CERN was indirectly a critical enabler of the Web, in the sense that Berners-Lee's superiors gave their tacit approval by allowing him to pursue his obsession – by giving him the freedom he needed in order to innovate. And the laboratory also provided an environment in which Robert Cailliau's formidable organizational and political skills could be fruitfully deployed in support of Berners-Lee's technical virtuosity.

But there is a difference between tacit approval and positive support. 'What we hoped for,' Berners-Lee wrote,

> was that someone would say, 'Wow! This is going to be the cornerstone of high-energy physics communications! It will bind the entire community together in the next ten years. Here

are four programmers to work on the project and here's your liaison with Management Information Systems. Anything else you need, just tell us.' But it didn't happen.[12]

Faced with the difficulty of obtaining internal CERN resources to write the software needed to get the functionality of the NeXT versions of the web software onto PCs, Apple Macintoshes and Unix machines, Berners-Lee and Cailliau decided that the only thing to be done was to harness the collective IQ of the Internet. So on 1 August 1991 they released three things – the WorldWideWeb code for NeXT, the line-mode browser, and the basic server for any machine – onto the Internet. Berners-Lee posted a notice on several Internet newsgroups (i.e. online discussion groups), chief among them alt.hypertext – the forum for hypertext enthusiasts. And having thus lit the fuse, he withdrew to a safe distance and waited for the explosion.

And now?

Nobody knows how big the World Wide Web is now, but as I write (in 2011) estimates of the size of the indexed web – that's the part that the major search engines reach – are hovering somewhere between 20 and 40 billion pages.[13] That's probably just the tip of the iceberg – below the indexed web there is the 'deep' web (i.e. pages lying behind organizational firewalls) which people think could be anywhere between 400 and 750 times the size of the visible tip.

Just think about those numbers for a moment. This is an avalanche of data and information that is beyond imagining. It has accumulated in the space of less than two decades. It has transformed our communications environment. Life without the Web is now unimaginable.

And yet this transformation was triggered by a single individual, working mostly alone on an unofficial project within a large organization which was, for the most part, blissfully unaware of what he was up to. He was able to trigger the avalanche without having to ask anyone's permission. Tim Berners-Lee had a great idea that could be implemented using data-packets. He released those packets onto the Net and the network then did what it was designed to do. In the process it propelled Tim Berners-Lee into Gutenberg's shoes. And took the rest of us completely by surprise.

Napster: the Celestial Jukebox

In his seminal meditation on the nature and significance of Open Source software, *The Cathedral and the Bazaar*, Eric Raymond cites as the 'first lesson' that 'Every good work of software starts by scratching a developer's personal itch'.[14] In the case of Tim Berners-Lee, the itch was his long-standing obsession with finding a way of storing and retrieving information. Ten years after the creator of the Web submitted his project proposal to his boss, an unruly teenager across the Atlantic in Massachusetts was also scratching the itch that had been bothering him. His name was Shawn Fanning, and his personal obsession concerned online music.

Fanning was born in 1980 in Brockton, Massachusetts, a town whose main claim to fame up to that point was that it had been the birthplace of Rocky Marciano, who was the heavyweight boxing champion of the world from 1952 until 1956, when he quit as the only heavyweight champion in history to retire having won every fight in his professional career.

Fanning was born into modest circumstances and, at least according to some reports[15], had a somewhat troubled childhood. He was, however, fortunate in having an uncle, John Fanning, who took an interest in the lad and encouraged him to get good enough high-school grades to be admitted to Boston's Northeastern University. His nephew turned out to be a 'bright, if not very diligent' student, much given to video gaming and to **Internet Relay Chat** (IRC), the first truly global text-messaging and **synchronous** conferencing system which served as an online meeting-place for geeks, hackers and traders of illicitly copied information goods of various descriptions.

As luck would have it, John Fanning was also an early Internet entrepreneur. With two other colleagues, he had built the world's first Internet chess server in the 1980s. This led to the founding of Chess.net and, later, of NetGames. Noticing that his nephew had developed good programming skills at college, John bought him a PC, offered him a part-time job and found that he thrived under the pressure.

What really interested the young Fanning was not gaming, however, but music. And by the time he got into computing in the mid-1990s there was a vast amount of music on the Internet. In fact there had been lots of music on the Internet even in the 1980s. The problem was that it was difficult to locate; and even when you had found it, it was also difficult to acquire. These two difficulties combined to form the 'itch' that Shawn Fanning set out to scratch.

A brief (well, compressed) history of digital audio

At this point, we need to have a brief historical digression. If you're new to this stuff, it may have come as a shock to learn that music files were being exchanged across the Internet as far back as the dark ages of the 1980s. So here's a brief outline of the story.

In its 'original' form (i.e. the form in which it is made by performers), music is made up of pressure waves (which we call sound waves) that pass through the air and impinge upon our ear-drums. Up to 1981, the only way we had of copying and relaying these pressure waves was by faithfully copying the patterns they made. That's why, if you examine the grooves in a vinyl LP under a magnifying glass, you will see that the grooves follow a wavy pattern.

But by the end of the 1970s an industrial partnership of Phillips and Sony had developed a way of digitizing pressure waves by sampling the amplitude of the wave many thousands of times a second, and registering the size of each sampled amplitude in binary form, that is to say as a string of ones and zeroes.

The basic idea of sampling in this way is illustrated in Figure 3.3

It turns out that if you take many thousands of samples every second, you can get a pretty good digital representation of the original, analogue, waveforms. For CD quality audio, the sampling frequency adopted was 44.1kHz, which is geek-speak for 44,100 samples per second. At that rate you have to be a real audiophile to be able to distinguish between the original – analogue – audio signal and its digitized representation.

In 1981 Phillips and Sony unleashed this new technology on the world. They called it compact disc or CD. The first players designed for the new format came on the market in 1982. So from that moment onwards you could say that all music was digital.

The recorded music industry took to this new technology with great enthusiasm – for obvious reasons. In the first place, it offered the industry a way of re-selling its back-catalogue:

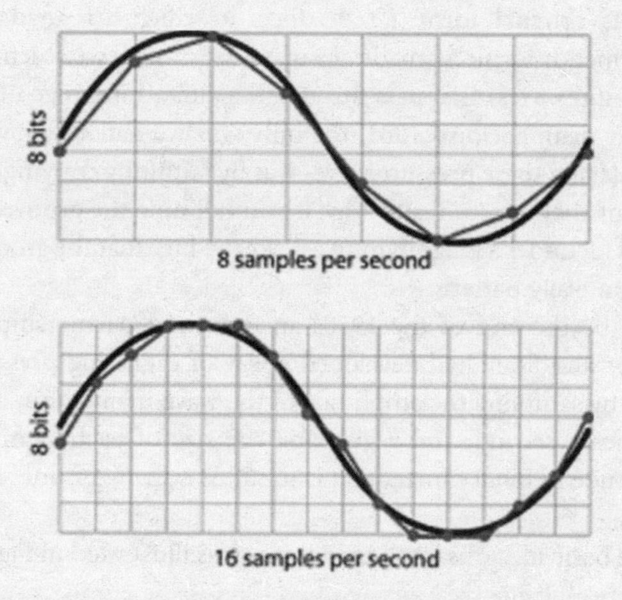

Figure 3.3: Sampling a sine wave. Note how the accuracy of the sampled wave (designated by the straight lines joining the dots) improves by increasing the frequency of sampling.

old analogue recordings could be 'digitally remastered' and re-issued as 'new' products. Secondly, the price level set for CDs was considerably higher than the levels which vinyl records were fetching as the market for them became saturated and went into decline – so profit margins were higher. So, all in all, the CD format gave a new lease of life to the music business.

Once music went digital, the way the industry operated could be characterized like this: At one end of the business there were the record labels who signed up bands and musicians. These artists went into recording studios and generated pressure waves which were converted into **bitstreams** by the sampling process outlined above.

At the other end of the chain were consumers with expensive devices (CD players) which could take bitstreams and convert them back into pressure waves that could be heard by the human ear. The problem was how to transport the bitstreams from studio to living room. The solution was to manufacture plastic discs onto which the bitstreams could be impressed. The discs were then labelled and put into clear plastic boxes (called, for some unfathomable reason, 'jewel cases'), which were packed in larger batches in cardboard boxes which were then loaded onto trucks and shipped to warehouses.

Later, in response to orders from retailers, the boxes of discs would be loaded onto more trucks and transported to retailers, who would then take the discs out of their jewel cases and store them at the back of the shop, leaving the empty cases to be browsed by customers. When someone purchased a disc, it was reunited with its case and given to the customer, who then took it home and inserted it in his or her CD player, which immediately decoded the bitstream and flooded the room with the desired audio signals. It was, as Nicholas Negroponte memorably observed,[16] a process of 'shipping atoms to ship bits' – and in 1982 it was a reasonable (if expensive and inefficient) way of transporting bitstreams from recording studio to living room.

For the record companies, the CD was a wonderful example of industry-friendly innovation because it offered a way of selling the same kind of stuff at a higher price (and with correspondingly higher margins) while not requiring any radical changes in their time-honoured customs and practices. In the old days, they made and sold vinyl discs; now they sold digital discs. Everything else – including the ways executives were incentivized – remained more or less the same. No disruption. No pain. Just more profit.

There was, however, one very important difference between CDs and their analogue, vinyl predecessors. Copying in an analogue world is a degenerative process, in that every copy inevitably has imperfections which make it inferior to the original from which it's been made. So while a copy of an original tape-recording might be of acceptable quality, a copy of a copy of a copy of a copy would be markedly inferior. Digital technology, however, doesn't suffer from this degenerative problem: every copy – even one made at the tenth or hundredth remove – is a perfect representation of the original. So once music went digital in the early 1980s, it also became infinitely copyable.

On 1 January 1983, just under a year after the appearance of the CD in record stores across the world, the Internet as we know it today (the network based on the TCP/IP family of protocols) was switched on. Although most people in the record industry – and in most other industries too, come to think of it – were blissfully unaware of the event, what it meant was that a global machine designed to transport bits from one place to another had arrived on the scene. The fact that those bits could include files of digitized music did not escape geeks, though, and soon tracks taken from CDs began appearing on the hard drives of computers connected to the network.

Initially, this didn't appear to pose much of a threat to the music industry – for two reasons. The first is that the files were huge, at least by the data-storage and bandwidth standards of the 1980s. Digitizing music to CD quality generated about 10 megabytes (MB) of data per minute, which meant that a three-minute music track took up more than 30 MB, and an album of 15 tracks 450 MB. (The maximum capacity of a CD was 700 MB.) Given that a '**broadband**' Internet connection in those days was deemed to have a throughput of perhaps 2 megabits (i.e. 250,000 bytes) per second, shifting music files around on any scale would have been quite a palaver.

The other frictional effect on file-sharing was that the technology for moving files around was pretty arcane. Essentially it involved using File Transport Protocol (FTP) one of the oldest **protocols** on the ARPAnet. Files were stored deep in hierarchically organized directories (i.e. folders) on networked computers, and in order to fetch one you had to (a) know where it was located, (b) have the requisite permissions to access the directory in question and (c) know how to use FTP in order to fetch it.

These two factors – the fact that very few people had access to the necessary storage and bandwidth facilities for handling large files, and that a measure of technical expertise and access was needed to share such files – meant that during the early years of the Internet's existence the swapping of music files was very much a minority sport. So I suppose it's not surprising that when the music industry thought about the Net in those days – to the extent that its executives thought about it at all – the network didn't look like much of a threat.

But that was only because they didn't know what was coming.

Enter MP3

As it happens, the large size of digitized music files wasn't just a challenge to the computing industry in the 1980s. It also troubled telephone companies which were likewise anxious to find ways of improving the quality of audio transmitted over their networks. (If you've ever listened to music relayed over a standard telephone landline you will know the problem.) The movie studios were asking similar questions in relation to digitized video files (which were orders of magnitude bigger than audio files.) Was there a way of transmitting less data while retaining perceived quality? Or, to put it in a more technical way: was there a way of compressing audio signals without compromising listener satisfaction?

As it happened, academics at the University of Erlangen-Nuremberg in Germany and researchers at the nearby Fraunhofer Institute for Integrated Circuits, a prestigious R&D lab which does contract research and development for industry and government, were asking themselves similar questions. One of them, Professor Dieter Seitzer, embarked on a project, funded by the European Union, to find a way of shrinking video files to acceptable sizes. Thus was born the Motion Pictures Expert Group or MPEG. As work on video compression proceeded, one of Seitzer's students, Karlheinz Brandenburg, embarked on his PhD research, but focusing purely on audio. Could one dispense, he asked, with parts of the signal without one's listeners noticing the loss – and thereby reduce the amount of digital data that would need to be transmitted?

The answer was 'yes' – providing one paid attention to how humans receive and process audio signals. In the first place, there's the physics of the ear. Most of us can hear frequencies only in the range between 20 to 20,000Hz (i.e. cycles per second). Our ears are most sensitive between 2KHz and 5KHz, and can detect changes between frequencies in increments of 2Hz – which is the effective 'resolution' of hearing. But as we get older our ability to hear high-frequency sound declines, to the point where most adults have trouble hearing anything above 16KHz. What this means to an engineer is that one can dispense with any data relating to sounds below 19Hz and above 20,000Hz without most people noticing.

All of this was well known when Brandenburg started work on his dissertation. His work involved re-examining and building on what was known about how the human brain *processes* auditory information – in other words what happens to the audio signal as it is passes from our ears to our brains. This is the field of *psychoacoustics*[17], and among its discoveries is that there exist 'audio illusions' analogous to the optical illusions produced by the way the brain processes visual stimuli. An example is the Haas effect i.e. the phenomenon that two identical sounds arriving within 30–40 milliseconds of each other from different directions are perceived as a single sound coming from the direction of the first. This is sometimes used in public address systems at concerts to give the illusion that the sound is coming from centre stage even though the loudspeakers are located to the side. Another set of audio illusions is created by 'masking' – the process of covering one frequency with another close (but not identical) one. As an example, suppose two sounds with similar frequencies – say 100Hz and 110Hz – but with different volume levels arrive at your ear at the same time. If played by itself, the weaker sound is perfectly audible, but you will only hear the stronger one if both are played simultaneously.

Building on these insights, Brandenburg and colleagues in the Fraunhofer Institute wrote an **algorithm** – a computer procedure – that could shrink CD files to between a tenth and a twelfth of their normal size, so a three-minute music file could be reduced to 3MB without much perceptible loss in quality. In 1991 they offered their solution to the audio-compression problem to the International Standards Organization (ISO) under the geeky label of 'MPEG-1 Audio Layer 3' – rapidly abbreviated to MP3.[18] It was approved as an ISO standard in 1993.

In July 1994, the Fraunhofer Institute released the first MP3 software encoder – i.e. a computer program that could take CD files and compress them using the MP3 filter. In 1995 the organization decided to use the suffix '.mp3' to designate files that had been compressed using the technique, and also filed several patents protecting the work that its engineers had put into the encoding procedures that underpinned the ISO standard.[19]

By the mid-1990s, therefore, a technology existed for shrinking music files to a manageable size. But most of the world remained blissfully unaware of the potential of this encoding technology. In 1997, however, Tomislav Uzelac, a Croatian programmer from the University of Zagreb, wrote a program that could play music encoded in the MP3 format. It was called AMP and ran on the Unix **operating system**. Later that year a talented drop-out from the University of Utah named Justin Frankel ported (i.e translated) it to run under the operating system that most people used – Microsoft Windows. Frankel called his program WinAmp and released it on the Net as *shareware* – if users liked it then they could send him $10. So many did that this 19-year-old suddenly found himself earning thousands of dollars a month. It was clear that something was up.

For many of its users, the first encounter with Frankel's program was a life-changing, paradigm-shifting experience. This is how it worked: you inserted a music album in the CD-ROM slot in your PC, WinAmp detected it and then began pulling each track off the disc, running it through the program's MP3 encoder and storing the resulting, compressed track on your hard drive. And at that moment, at least for some music lovers, the world changed. Before WinAmp, a music collection was a stack – or a shelf – of physical discs in your living room. After WinAmp, it was a set of files on your hard drive. Before WinAmp a 'music centre' was an assembly of audio components – turntable, CD player, amplifier; after WinAmp it was your PC hooked up to an amplifier. And although hard-drive space was relatively constrained and expensive in the 1990s, MP3 compression was so effective that one could fit an entire music collection (not to mention those of your family and friends) on a modest PC or laptop. Better still, if your PC had a drive that could write as well as read CDs, then you could go a step further – by producing compilation CDs which you could share with others. A new vocabulary emerged to describe the process: 'rip, mix and burn'.[20]

Even so, sharing music files at this stage still involved 'shipping atoms to ship bits', in the sense that in order to share MP3 files you had to buy blank 'writeable' CDs and physically distribute them to your friends. If anyone in the record industry was paying attention to what WinAmp was making possible, they could have consoled themselves with the thought that what was happening was not all that different from what had gone on in the good ol' analogue days when music fans would make compilations of their favourite tracks on cassette tapes and distribute them to their friends. Sure, it was irritating, but not a mortal challenge to a cosy and well-understood business model. And it was limited by the physical restrictions of the real world – the cost of raw materials and the difficulty of distributing discs.

Here endeth the digression. We're now ready to return to Shawn Fanning and his personal itch.

Napster

By the time Shawn Fanning arrived on the scene in the late 1990s, millions of music lovers across the world had got into the habit of keeping much, if not all, of their music collections as MP3 files on the hard drives of their personal computers. The trouble was that there was no organized way of searching for music held on people's PCs. You couldn't use a standard search engine, because that would only search web pages, not people's PCs. And even if people had uploaded MP3 files to the Web, more often than not a search engine would return links that were broken – the pages no longer existed or the MP3s had been deleted for various reasons. And beyond the usual social mechanisms of exchanges between friends, colleagues and classmates using specially burned CDs or email attachments, there was no established way of sharing MP3 files because the PCs of the day were not set up to act as servers.

Remember that this was a time when broadband connections were relatively scarce: most people still accessed the Net via slow dial-up connections. Worse still, their computers lacked permanent Internet Protocol (**IP**) addresses; on dialing in to an **Internet Service Provider** (ISP) a PC would be assigned a temporary address for the duration of the connection; next time the machine hooked up to the Net it might be assigned a different address from the batch held by the ISP. This meant that if I went looking for your computer on the Internet, it probably didn't have the same IP address as it had yesterday. The PCs with these non-persistent connections ran under relatively primitive operating systems like Windows 95.

What this meant in practical terms is that most domestic PCs were not capable of acting as *servers* of content (e.g. music tracks); they were designed to function mainly as relatively dumb *clients* which requested services from powerful, **multi-tasking** servers permanently connected to the Net via broadband links.

Shawn Fanning's genius was that he found a way of overcoming these crippling deficiencies. With two collaborators whom he had met online – Jordan Ritter and Sean Parker – he created a software system with two components. The first was a client – a small program that anyone could download and install on their PC. The second was special software that ran on a permanently connected server. He called his system Napster[21], which according to some sources was a nickname he had been given at school.

If you had Napster installed on your PC, here's how it worked. When you connected to the Net, the Napster client would activate, check the **IP address** assigned to your PC for the session, and make a list of the MP3 files currently in the 'shared' folder on the hard drive. It would then contact the Napster server and say, effectively, 'I'm online at [this IP address] and here's a list of the files in the shared folder'. The Napster server would log this information and store it in its database.

Meanwhile, elsewhere on the Net, another Napster user would type a track name into the search box that came with the client. The server would analyse the query, search its database and then return a list of the PCs that were currently online and had the desired track in their 'shared' folders. (Figure 3.4.) Clicking on one of the entries in the list would then initiate a direct file-transfer between the two machines.

In technical terms, Fanning's Napster system was remarkably neat. He had found a way of enabling the second-class citizens of cyberspace – PCs with only dial-up connections – to locate one another and exchange files: to function, in essence, as servers despite not having persistent IP addresses or multitasking operating systems. The system was conceptually elegant and clear, and the software was well-written and easy to grasp and use. For a young, hitherto-unknown programmer, it was an astonishing achievement.

In June 1999, Fanning released Napster onto the Net. It spread like wildfire, especially on college campuses across the US, where students were beginning to enjoy broadband access as one of the privileges of university life.

It's hard to get definitive statistics for the number of users Napster attracted during its brief life. Comscore data suggests that usage peaked in early 2001 at around 28 million, but other estimates are as high as 70 million.

In a way, what was more significant was not the number of users, but the astonishing variety of the music that the service made available. In its heyday, it didn't seem fanciful to believe that virtually every piece of popular music that had ever been recorded – including the vast proportion of it that was effectively unobtainable because it was 'out of print' – was suddenly available. Shawn Fanning had built what one writer called 'the

Celestial Jukebox'.[22] He had given the world a glimpse of what the Net could do.

The trouble was, of course, that most of the tracks that Napster users shared were copyrighted works. Fanning and his colleagues persuaded themselves that since their service did not host any music on its servers they might be able to escape litigation. In that they were mistaken, and in the summer of 2001 the legal campaign waged by the record industry against the service succeeded. A judge in California ordered the service to shut down.[23]

The trouble for the record industry was that by that stage the genie was out of the bottle. An entire generation of teenage music lovers had been exposed to the charms (and benefits) of file-sharing. And even as Napster and the record labels were slogging it out in court, new kinds of file-sharing architectures had emerged. One weakness of Napster was that it needed a 'mother ship' – the server which logged the IP addresses and MP3 files of its online subscribers; such a server provided an identifiable focal point for legal action by lawyers claiming infringement of copyright. The next generation of file-sharing software – systems like Gnutella, Limewire, Freenet, BitTorrent and so on – did not suffer from this disability. Illicit file-sharing continued to thrive – and still thrives.

Napster was a world-changing surprise that was enabled by the Net. Shawn Fanning had an ingenious idea which he implemented in software. He then released it onto the Internet without having to seek anyone's permission, and the network did the rest. For music fans, Napster was a pleasant – nay delightful – surprise. For the music industry it was the most unpleasant shock imaginable, and one that triggered a nightmare from which it is still struggling to recover.

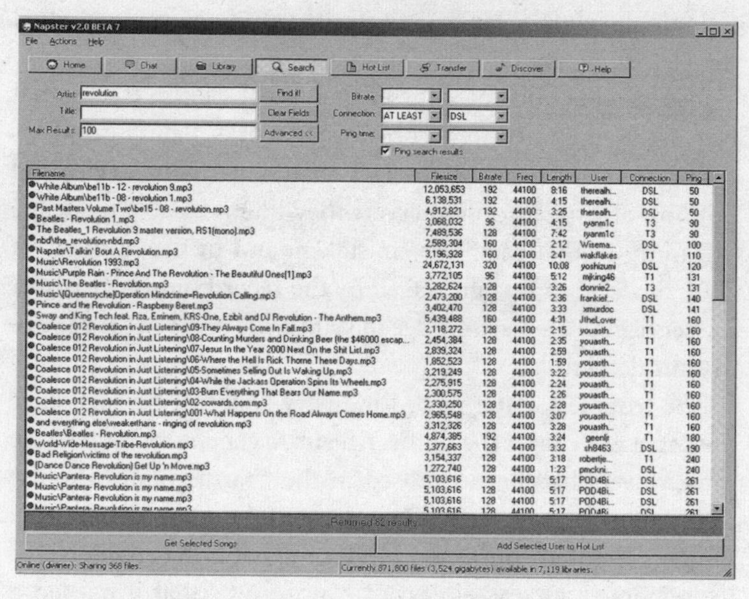

Figure 3.4: screenshot of a Napster search for 'revolution'

The greatest irony of the story is that it need not have happened like this. After the launch of the compact disc, the recording industry needed an efficient and cost-effective way of shipping its bitstreams to customers. Its original method of doing this – via plastic discs – was expensive and inefficient. Upwards of half the cost of a CD was accounted for by distribution costs. The Internet offered a way of shipping those bits in an efficient and almost cost-free way – eliminating fifty per cent of the product's cost at a stroke.

In the early years of the CD (the 1980s), the music industry would have been justified in rejecting the Internet as a way of distributing its product. At that stage, the Net was effectively confined to the worlds of academia and research. But from the 1990s onwards – and especially as domestic broadband connec-

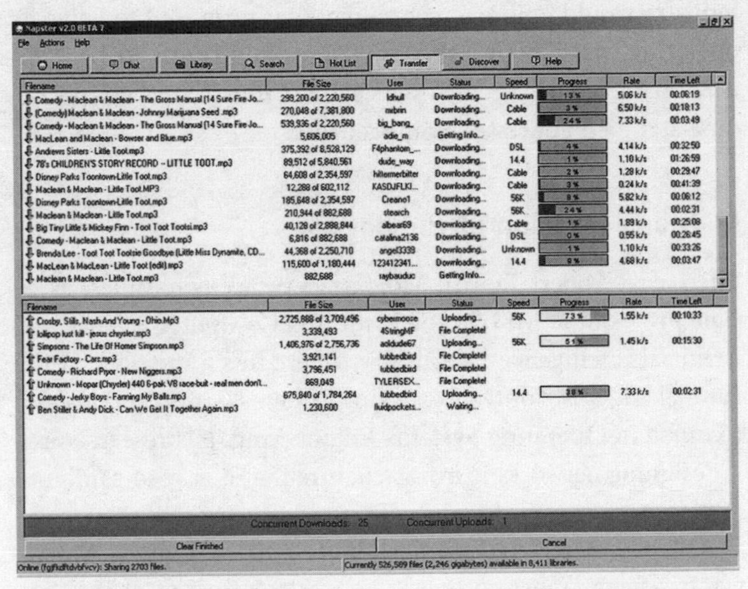

Figure 3.5: Screen shot of a Napster session in progress. The top pane of the window shows the status of files that the user is downloading to his/her machine. The bottom pane shows files that are in the process of being uploaded from the shared folder on the user's computer.

tions increased – the network came to look like a viable distribution system.

And yet the record industry effectively ignored it. Not only that: it persisted with a distribution system whose economics meant denying customers what they wanted. CD technology meant that single tracks were economically unviable and so the industry tried to force its customers to buy albums. Yet the demand for 'singles' – tracks – never went away, as the world discovered when Napster provided an easy way to get them.

MBA students of the future will pore over the file-sharing phenomenon as a case study in corporate psychosis. What other industry would ignore a technology that promised to halve its operating costs? What other industry would turn its back on what many of its best customers wanted most? Answers, please, on a stamped addressed compact disc.

Malware: Invasion of the botnets

These days Robert Tappan Morris is a perfectly harmless associate professor at MIT, in the Institute's Computer Science and Artificial Intelligence Laboratory, where he's a member of the Parallel and Distributed Operating Systems Group and teaches a course in Operating Systems Engineering. His research area is designing networking infrastructure that is easy to configure and control.

As I say, perfectly harmless. But in 1988, Robert Morris was briefly the most notorious man in cyberspace. At the time he was a 23-year-old graduate student at Cornell University, in Ithaca, NY, and he wanted to find out how many computers were connected to the Internet. To do this he wrote a small program that exploited loopholes in Internet mail protocols and the Unix operating system so that it could covertly install itself on computers connected to the network. Once installed it would send a 'present and correct' message back to Morris, enabling him to register another networked machine. It would then look through any electronic address books it found on its unwitting host in search of other networked machines with which the machine had dealings, after which it would then replicate itself on them. Its purpose was simply to transmit a copy of itself to the machines, where it would run alongside the other software on the machine and repeat the cycle.

In designing his 'worm' (for that is what it came to be called), Morris made a decision which had unintended – and serious – consequences. Because it was important (from his research perspective) to avoid double-counting of computers, the program checked before invading a new machine to see if it had already been infected. But this offered a way of stopping the count in its tracks: if a network administrator on a target machine had figured out what was happening, he could simply fool the Morris worm by having his machine answer 'yes' to the query, in which case the worm would look elsewhere. Morris's solution to this was to program the worm to replicate itself one in seven times, even if the answer to its initial query was affirmative. It turned out that this level of replication was excessive.

On the evening of 2 November 1988, many computers connected to the Internet began behaving strangely. As Jonathan Zittrain describes it:

> Unusual documents appeared in the depths of their file systems, and their system logs recorded activities unrelated to anything the computers' regular users were doing. The computers also started to slow down. An inventory of the running code on the machines showed a number of rogue programs demanding processor time. Concerned administrators terminated these foreign programs, but they reappeared and then multiplied. Within minutes, some computers started running so slowly that their keepers were unable to investigate further. The machines were too busy attending to the wishes of the mysterious software.[24]

System Administrators (*sysadmins* in computerspeak) quickly twigged that Morris's 'unauthorized' code was spreading across the Internet from one machine to another and, working together – often via phone conferencing – they collectively figured out how to disable the infecting software.[25] By the end of the day it was reckoned that between five and ten per cent of all machines on the Internet had been infected. Dr Cerf's surprise-generating machine had done its stuff – and this time the surprise was distinctly unpleasant.

Although Morris's worm had relatively innocent origins, it wasn't the first example of unauthorized software on the network. In 1971, Bob Thomas, an engineer working for Bolt Beranek and Newman, the Boston company that had the contract to build the ARPAnet, released the first known computer virus – which he called Creeper – onto the network. It was an experimental, self-replicating program that infected DEC PDP-10 minicomputers and did no actual harm but merely displayed a cheeky message: 'I'm the creeper, catch me if you can!' Someone wrote a program to detect and delete it, which was – inevitably – called The Reaper.

Morris's worm and Thomas's Creeper were harbingers of things to come. Ever since the advent of the personal computer in the mid- to late 1970s, programmers had been experimenting with ways of writing and distributing malicious programs. In 1982, for example, a Pennsylvanian teenager named Rich Skrenta created a computer virus aimed at the Apple II computer which was, at the time, the most popular personal computer in upmarket US households.

The Apple machine did not have a hard disk, and so one had to insert a floppy disk containing the operating system in order to get it going. Once the machine was powered on, the first thing it did was to read a special sector of the disk –

called the *boot sector* – and load its contents into memory. This then told the machine what to do next, enabling it to find and load the next part of the operating system – to, as it were, pull itself up by its bootstraps; hence the term *booting up*.[26]

Skrenta wrote a program – subsequently called a virus – that covertly altered the boot sector so that the machine would load the virus into memory and then carry out its instructions, which in this case were to display some doggerel on the screen. It read:

> Elk Cloner: The program with a personality
> It will get on all your disks
> It will infiltrate your chips
> Yes, it's Cloner!
> It will stick to you like glue
> It will modify RAM too
> Send in the Cloner!

Computer viruses are so-called because, like biological viruses, they piggyback on other organisms. In the computer context these are programs that people use as a matter of course – for example Microsoft Word. Every time the infected file is run the virus runs too. The 1980s saw an explosion in virus-writing, but the problem was contained by the fact that most personal computers in that decade were not networked and so viruses could only be transmitted by *physical* means – for example by people swopping floppy disks containing infected files.[27]

All of this changed in the 1990s once PCs began to be connected to the Net in large numbers. Suddenly we entered a world in which millions of computers – most of them operated by technically naïve users with little or no understanding of security issues – were connected to a global network. And this in turn created a perfect environment for the rapid dissemination of malicious code ('malware') of all descriptions – from worms which used the network to propagate themselves; through viruses that piggybacked on popular software products and email messages; to 'Trojan horses' – programs masquerading as a game or some other kind of intriguing software but which, when activated, do something else – like erasing your hard drive or monopolizing the screen with a graphic that will not go away.

In the early days of the malware boom, most of the viruses and worms that spread across the Net had innocuous payloads. The MyDoom worm of 2004, for example, targeted computers running Microsoft Windows, spread via email and affected millions of computers. But although it obviously 'cost' millions of dollars in lost productivity as people laboured to disinfect their computers, it did not tamper with data and was programmed to stop spreading at a set time.[28]

Looking back, the era of MyDoom looks like the Age of Innocence. The malware of the time seems largely to have been the product of vandals or pranksters. As Zittrain put it,

> Programs to trick users into installing them, or to bypass users entirely and just sneak onto the machine, were written only for fun or curiosity, just like the Morris worm. There was no reason for substantial financial resources to be invested in their creation, or in their virulence once created. Bad code was more like graffiti than illegal drugs. Graffiti is comparatively easier to combat because there are no economic incentives for its creation. The demand for illegal drugs creates markets that attract sophisticated criminal syndicates.[29]

Or, to put it another way, early malware might have been irritating or infuriating, but it wasn't dangerous to (online) society for one simple reason: it lacked a business model.

Two things combined to change that. The first, for some reason, went under the same name as a famous type of processed meat: it was called **spam**[30]. The second was the growth in broadband access to the Net.

Spam, spam, inglorious spam

Spam is unsolicited 'junk' email. As it happens, we know more or less exactly when it started. And it was a surprisingly long time ago – on 3 May 1978 to be precise. Its hapless author was a computer salesman named Gary Thuerk who worked for the Digital Equipment Corporation (DEC)[31], then a leading minicomputer manufacturer. DEC was located in Massachusetts on the East Coast and had developed a new line of quasi-mainframe computers, the DEC-20 range. In preparation for a product launch in California, Thuerk emailed everyone on the West Coast ARPAnet email list, and thereby earned his place in history. His message began:

> DIGITAL WILL BE GIVING A PRODUCT PRESENTATION OF THE NEWEST MEMBERS OF THE DECSYSTEM-20 FAMILY; THE DECSYSTEM-2020, 2020T, 2060, AND 2060T. THE DECSYSTEM-20 FAMILY OF COMPUTERS HAS EVOLVED FROM THE TENEX OPERATING SYSTEM AND THE DECSYSTEM-10 <PDP-10> COMPUTER ARCHITECTURE. BOTH THE DECSYSTEM-2060T AND 2020T OFFER FULL ARPANET SUPPORT UNDER THE TOPS-20 OPERATING SYSTEM.[32]

How unsolicited junk mail came to be known as 'spam' is unclear. Brad Templeton attributes it to the celebrated *Monty Python* sketch in which a restaurant serves all its food with lots of spam, and the waitress repeats the word several times in describing how much spam is in the items. When she does this, a group of Vikings in the corner begin singing: 'Spam, spam, spam, spam, spam, spam, spam, spam, lovely spam! Wonderful spam!'[33]

The first junk email message that attracted widespread odium was composed by Canter and Siegel, a law firm based in Phoenix, Arizona, advertising their services to people applying for a Green Card (a card confirming US residence status). It was sent in April 1994. In itself this might not have provoked outrage, but the firm hired a programmer to write a script that posted the ad to every single forum within the vast **Usenet** system of Internet discussion groups. Within the Usenet culture, which still largely enshrined the original, non-commercial ethos of the Net, the Canter & Siegel message was akin to an offence against nature – an intrusion of commerce into a space that had hitherto been untainted by it. It turned out, however to be a foretaste of worse things to come.

Much worse. By 2009 spam was accounting for an average of 85.5 per cent of global email traffic.[34] And 0.3 per cent of spam messages included malicious attachments – mostly small programs that did nasty things to your computer, or turned it into a relay station for doing the same to others. By the end of the first decade of the twenty-first century it looked as though the Internet was being choked by avalanches of junk – unsolicited advertisements for sex aids, counterfeit products, herbal 'remedies', stock-market tips, pornography and more.

Technology-savvy users gradually learned to avoid the worst effects of this by using spam-blocking services and filters of various kinds, but many non-technical users became inured to having their mailboxes (and their bandwidth) choked by the tide of electronic junk. And from time to time they must have wondered why it has seemed impossible to do anything about it.

After all, if I send you an email, my address is included in the message – and even if it isn't, the communication can ultimately be traced back to me by someone with the right forensic tools. So why can't spammers be identified and brought to justice?

The answer goes back to the philosophy of openness embodied in the Net's architecture. The software protocol that handles email – Simple Message Transfer Protocol (SMTP) – emerged in the early 1980s in the engineering community that was designing the Internet. This was a community of researchers who knew and trusted one another, and so the protocol they designed for handling email messages did not have a provision for authenticating senders – i.e. for forcing them to establish that they were who they claimed to be. It simply took mail messages at their face value and passed them on to their destinations. This worked fine in the beginning, because those who used the network were, in the main, legitimate and trustworthy. But once the Internet became a mass phenomenon, the lack of authentication in SMTP was exploited by spammers who were able to 'spoof' the system by providing return addresses that were fraudulent – either because they were non-existent, or because they are legitimate but belonged to someone else.

Why is there so much spam? Simple: it's profitable. This may seem odd to you, given the crass implausibility of the products and services offered by junk emails. Who would be stupid enough to fall for such nonsense? The truth is that only a tiny proportion of those who receive spam take the bait, but that tiny percentage is enough to ensure a healthy return on the spammers' investment. Andrew Leung, an analyst employed by a Canadian telecoms firm, estimates that spammers can make 'a lucrative living' if as few as fifty people in a million respond to a junk mail.[35]

'Spam is all about economics,' writes the security expert, Bruce Schneier. 'When sending junk mail costs a dollar in paper, list rental, and postage, a marketer needs a reasonable conversion rate to make the campaign worthwhile. When sending junk mail is almost free, a one in ten million conversion rate is acceptable'.[36]

What it comes down to is this: if you send 1,000 spam messages a day advertising fake Viagra at $20 a bottle and you have a response rate of 0.1 per cent then that means a revenue of $20 per day. But if you send 10 million messages and have the same response rate, then your revenues are $200,000 per day. The key to success in spamming, therefore, is volume.

But there is a downside to volume. If your organization is spewing out ten million emails a day, then someone – your ISP, for example – is going to notice, and is likely to inform the police about your activities, because spamming is nowadays illegal in many jurisdictions. So the challenge for the spammer is how to achieve volume without being noticed.

The requirements for effective spamming are:

- Covert but effective control of lots of computers, dispersed geographically across the world, each of which can be used as a relay station for spam.
- These computers should not be owned by, or traceable to, the spammer. Their hapless owners should have no idea of what their computers are up to on the Net.
- Ideally, these relay stations can be switched on and off at will, so that it's difficult for the authorities to pin-point the source of the junk emissions.
- But when they're active, the relay stations should be able to spew out a vast number of emails in a short time.

Which is where broadband comes in.

The width of a band

Just as every pipe in a plumbing system has a maximum throughput (measured, say, in litres per minute), every electronic communications channel has a maximum capacity which is called its **bandwidth**. In analogue systems this is a measure of the range of frequencies that the channel – telephone line, cable, fibre-optic strand, etc. – can handle. In digital communications, bandwidth is measured in terms of the number of bits a channel can transmit in a second.

In the early days of computing, the main channel for communication between machines was the ordinary copper cable of the telephone line. This had an exceedingly low bandwidth – 1,200 bits per second (bps) was not unusual at the beginning, though by the 1990s various technical advances (mainly in data-compression techniques) had enabled computer users to get speeds of up to 56,000 bps (56 kilobits – kb – per second) on their dial-up phone lines. This sounds impressive, but you need to divide it by eight to get the capacity in *bytes* per second. So 56kbps translates into 7,000 bytes per second.

For applications like email – which are mostly about sending characters – such speeds are adequate. But once the Web went mainstream in the 1990s, and web pages became increasingly top heavy with data-intensive graphic images, the 56kbps dial-up channel began to seem frustratingly narrow. Browsing the nascent web came to feel like filling a domestic water cistern through a narrow-bore pipe. And the really data-intensive media like video were, effectively, beyond the reach of the domestic consumer.

The arrival of the Web spurred consumer demand for high-bandwidth – broadband – Internet connections. Initially, this was defined as anything offering speeds of 512kbps or more, and it was achieved by a number of technological advances: putting special equipment in telephone exchanges that effectively increased the throughput of phone lines to a maximum of eight million bits (megabits) per second; using TV cable lines to provide Internet connectivity on top of cable television services; and the deployment of fibre-optic cables which can handle throughputs of 1,000mbps (one gigabit per second) or more. The result of these developments has been a rapid increase in the penetration of broadband connections across the industrialized world. In the UK, for example, seventy-three per cent of all households have an Internet connection, and of those, seventy-one per cent are defined as 'broadband'.[37]

This sounds impressive, but in fact the UK is, relatively speaking, a laggard in this area. In its *Digital Britain* strategy document[38], published in 2009, the government set out a target of getting every connection in the UK up to a speed of 2mbps by 2012. Compare this with Japan and South Korea, where some *domestic* consumers have 1gbps connections; or with Finland (110mbps), Denmark, France, Iceland and Sweden (100mbps) or even the Netherlands, which has 60mbps.[39]

The implication of the widespread dissemination of broadband, however, is clear: all across the industrialized world there are millions of computers which have high-capacity, permanent connections to the Internet. This is great for their owners, who can now access sophisticated audio-visual and interactive media from the comfort of their own homes. But it has also proved a boon for the vandals and criminals of cyberspace. >

Botnets

A '**bot**' is a software robot – i.e. a program that is designed to do something mindless and repetitive under remote control. The program that a search engine uses to visit websites and trawl through their pages gathering data for its index is a benign example.

But bots can also have malign functions. They can, for example, be used as relay stations for spam. Or, even worse, they can be used to launch 'denial-of-service' (DOS) attacks on public websites. This is done by bombarding the target site with so many simultaneous electronic requests that its servers are overwhelmed, thereby making the site (and the service it offers) unavailable.

As with spam, volume and untraceability are the keys to successful DOS attacks. Millions of simultaneous 'pings' have to be generated by many different, geographically dispersed, computers so that the attack becomes a distributed DOS (or **DDOS**) attack. In order to do this, malware authors now assemble 'botnets' – networks of geographically dispersed PCs with broadband connections which have been secretly infiltrated by the **botnet** creator[40] using malware that has not been detected by their hapless owners. PCs that have been compromised in this way have effectively become 'sleepers' that can be woken up by the botnet controller and used to launch DDOS attacks or relay spam.

Today's Internet is riddled with botnets, some of which are very extensive. One source estimates that the Conficker worm, for example, has compromised more than ten million PCs, giving the botnet a spam capacity of ten billion messages per day.[41] The same source puts the size of the botnet created by the Storm worm at 85,000 zombie PCs, of which only 35,000 are used to relay three billion spam emails per day. (It's not clear what the remaining 50,000 PCs are used for, but DDOS attacks are clearly one possibility.)

If today's Internet is plagued by spam and malware, then it is a consequence of its open, permissive architecture. It turns out that it isn't only the good guys who are ingenious and inventive. But then we always knew that.

Second-order surprises

Wikipedia

Once upon a time, an encyclopedia was a shelf-full of bound, printed volumes. For most of my life, the gold standard in the encyclopedia business was the *Encyclopedia Britannica*, which in its fifteenth (1974) edition ran to thirty volumes and required its own book-case. A few weeks ago, while working on this chapter in the wonderful, book-lined Reading Room of the University Library in Cambridge, I found myself suddenly needing to check a date. In the normal course of events, I would have opened my laptop and gone to Google, but on a whim I decided to do something different. 'There must be a copy of *Britannica* here,' I thought, 'why don't I use that?'

So I left my laptop unopened and went looking. I found what I was seeking, sitting in a long serried row of brown volumes on a shelf in a corner of the Reading Room. And as I took one of them down I had one of those quiet moments of revelation that James Joyce called epiphanies. It was clear, first of all, that this edition of *Britannica* hadn't been used recently: indeed one could tell from the condition of the volumes that most of them had received very little use. But the biggest shock came from realizing how dated the entries were and how narrow the coverage was. And suddenly one realized how much the world of encyclopedic knowledge has changed. And how *Britannica* belongs to the past.

The prime agent of this change is a development that nobody (save perhaps Douglas Adams, author of *The Hitchhiker's Guide to the Galaxy*[42]) would have predicted. It is called Wikipedia – an online encyclopedia collectively produced and edited by 'amateurs' who have created what is effectively the greatest reference work the world has yet produced. I've just checked the Wikipedia site (www.wikipedia.org). It's reporting that the English language edition of the encyclopedia currently has 3.61 million articles; the German edition has 1.21 million; there are 1.09 million in the French edition, 0.79 million in Italian and roughly the same number in Polish. And there are editions in

Chinese, Japanese, Portuguese, Dutch, Russian, Swedish and even Esperanto.

As I write, Wikipedia has about 360 million readers and is the eighth most visited site on the Web, behind Google, Facebook, YouTube, Yahoo, Blogger.com, Baidu (the leading Chinese language search engine) and Windows Live (Microsoft's search engine).[43] All of these are Internet properties run by vast corporations with annual budgets running into billions of dollars. In contrast, the Wikimedia Foundation – Wikipedia's parent organization – had about forty employees in mid-2010[44] and an annual budget of less than a million dollars.

In January 2008, a group of researchers from Carnegie Mellon University and Xerox PARC attempted the first comprehensive mapping of the distribution of topics in the 2.5 million articles that the English edition of Wikipedia then contained.[45] This is what they found:

% of articles	Topic category	% change since 2006
30%	Culture & the arts	+210%
15%	People & self	+97%
14%	Geography & places	+52%
12%	Society & social sciences	+83%
11%	History & events	+143%
9%	Natural & physical sciences	+213%
4%	Technology & applied sciences	-6%
2%	Religion & belief systems	+38%
1%	Mathematics & logic	+146%
1%	Philosophy & thinking	+160%

Wikipedia's coverage is astonishing in breadth, depth, timeliness and – sometimes – triviality. There are, for example, detailed articles on TV reality shows, extensive entries on the Nintendo Pokemon video games franchise, and on the Barbie fashion doll created by the Mattel Corporation. At the same time, there are detailed, well-informed entries on the Internet Protocol Suite; the economist/philosopher Friedrich Hayek; the Bose-Einstein condensate; chaos theory; Jacques Lacan, the celebrated French psychoanalyst; Lloyd George; Ludwig Wittgenstein . . . The list is – or appears to be – endless; and if you search for a topic on which Wikipedia does not at present have an entry, you are presented with an invitation to create one. (*En passant*, searching for a topic that Wikipedia does *not* already cover is rapidly coming to be like searching for an English-language domain name that hasn't already been registered.)

The point is that anyone can create or edit a page. To most people, this would suggest a recipe for chaos. After all, as the *New Yorker*'s Stacy Schiff pointed out in an article published in 2006, Wikipedia is no more immune to human nature than any other Utopian project:

> Pettiness, idiocy, and vulgarity are regular features of the site. Nothing about high-minded collaboration guarantees accuracy, and open editing invites abuse. Senators and congressmen have been caught tampering with their entries; the entire House of Representatives has been banned from Wikipedia several times. (It is not subtle to change Senator Robert Byrd's age from eighty-eight to a hundred and eighty. It is subtler to sanitize one's voting record in order to distance

> oneself from an unpopular President, or to delete broken campaign promises.)[46]

Schiff was writing in 2006, when Wikipedia was five years old and often mired in controversy. This stemmed from various sources. First, there was the kind of visceral incredulity that comes from what the legal scholar James Boyle calls 'cultural agoraphobia', by which he means 'an asymmetry in our perception of risks and benefits in the world of intellectual property, the kind of property that lives on networks':

> We are very good at seeing the downsides and the dangers of open systems, open production processes [and of the] networks of openness. And let me be clear, those dangers and downsides are real. It's not that there aren't any. [But] we are not so good at seeing the benefits and the converse holds true for the closed system. It's easier for us to understand the benefits of closed systems and harder for us to see the downsides.[47]

Controversy was also stoked by sharp criticism from competitors (e.g. the online version of *Britannica*), by the derisive scepticism of sectors of the academic community (some of whom banned their students from citing Wikipedia in assignments) and by occasional discoveries or exposés of error or abuse by Wikipedia contributors.

The most notorious case of abuse came to light in May 2005 and concerned the anonymous posting of a hoax article about John Seigenthaler, a well-known American journalist. The article fabricated statements that Seigenthaler had been a suspect in the assassinations of US President John F. Kennedy and Attorney General Robert F. Kennedy. The 78-year-old Seigenthaler, who had been a friend and aide to Robert Kennedy, described the Wikipedia entry about him as 'Internet character assassination'. The hoax was not discovered and corrected for more than four months, and eventually led to a change in the rules under which Wikipedia operates by barring unregistered users from creating new articles.[48] The episode prompted the scientific journal *Nature* to publish in December 2005 a study comparing the accuracy of Wikipedia and the *Encyclopædia Britannica* in 42 hard-science-related articles. The Wikipedia articles studied were found to contain 4 serious errors and 162 factual errors, while *Britannica* also contained 4 serious errors and 123 factual errors. Referring to the Seigenthaler incident and several other controversies, the study concluded that 'such high-profile examples are the exception rather than the rule.'[49]

It is said that, according to the laws of aeronautics, the bumble-bee ought not to be able to fly. But fly it does. Much the same might be said of Wikipedia: it ought not to work, and yet it does. Over the years since the Seigenthaler incident, the site has grown not just in size but also in intellectual stature, to the point where it has become indispensable for many routine purposes and essential for some important ones. This is particularly evident whenever a major catastrophe strikes. For example on 6 April 2009 a large earthquake struck central Italy at 1.32 GMT. By 2.20 GMT there was a Wikipedia entry about it which was regularly updated during succeeding hours and days.[50] The same thing happened with the Indian Ocean tsunami of 26 December 2004 and the Japanese earthquake and tsunami of 11 March 2011. In these cases, where conventional news agencies struggle to cope, Wikipedia is turning out to be the best source of information and explanation.

However one looks at it, Wikipedia is an astonishing phenomenon. From the point of view of this chapter, three aspects of it stand out.

The first is the astonishing amount of collective human effort that it represents. In his book *Cognitive Surplus*, the cultural commentator Clay Shirky makes a stab at quantifying this. 'Suppose,' he writes,

> we consider the total amount of time people have spent on it as a kind of unit – every edit made to every article, and every argument about those edits, for every language that Wikipedia exists in. That would represent something like one hundred million hours of human thought . . . It's a back-of-the-envelope calculation, but it's the right order of magnitude.[51]

The intriguing thing about this massive collective investment of time and effort is that it is *voluntary*. People create and edit Wikipedia for motives that have nothing to do with the more conventional motivations of human action. This is what Yochai Benkler calls *social production*, creative activity that lies outside the marketplace[52], and in that sense is another instance of the rise of user-generated content that has been enabled by the Web (and which is discussed more fully in Chapters 4 and 8). But there is one important difference: most other kinds of user-generated content (for example, blogging, publishing photographs on Flickr or home-produced movies on YouTube) are mainly motivated by a desire for individual self-expression, whereas the effort that goes into Wikipedia is directed at a collective goal – that of creating the greatest work of reference that has ever existed.

The second thing that is fascinating about Wikipedia is why such an ostensibly uncontrolled and free enterprise has not degenerated into chaos. The answer lies in a combination of social conventions and technology. The site has remarkably few formal rules, but it does have a philosophical ethos that translates into norms that govern behaviour. This ethos includes: a commitment to transparency about editorial decision-making; a guiding principle that contributors and editors should strive to adopt a 'neutral point of view'; a commitment to the notion that there should be no 'gatekeepers' as far as content is concerned; and a pragmatic design philosophy that emphasizes the value of creating something that works *now* and assumes that problems can be dealt with as they arise.

Without the right technology, however, this philosophical ethos might have been aspirational rather than operational. What made it practical was the adoption of the Wiki software developed by an American programmer, Ward Cunningham, in 1994–95. Cunningham's great idea was to invent a way that enabled web pages – which had hitherto been read-only objects published on servers – to be edited, on the fly, by readers in their web browsers.[53] The adoption of this technology by Wikipedia's founders in 2001 was arguably the most important decision they made.

Why? Because it enabled the site to implement its ethos, as it were. Editorial transparency was ensured by the fact that the software automatically created and maintained a discussion page alongside every main page. This allowed people to explain their edits, and allowed those who disagreed to argue back – in public. Controversial edits that were made without any corresponding explanation on the discussion page could be reversed on the grounds that the absence of an explanation provided *prime facie* evidence for being suspicious of the changes. The discussion page provided a channel for debate that also introduced potential contributors to the social norms governing editing, and thus may have eased their transition from being readers to becoming editors.

Wiki software also provided an insurance policy against what many of the sceptics thought would be Wikipedia's greatest weaknesses – its vulnerability to vandalism. Because the program keeps track of every single change, and logs all activity on the site, it made it trivial to undo damage by simply reverting to the previous version of the defaced page. In that sense, the technology dramatically lowered the cost of dealing with vandalism and boosted the self-correcting capacity of the site.

In a way, though, the most interesting thing about Wikipedia is that it is a purely networked enterprise. That is to say, it's something that could not have come into being without the Internet, and it could not be sustained without it. Which is why it makes such an interesting case study for this book.

Facebook

In 2011, the two biggest countries in the world, as measured by population, were the People's Republic of China (with 1.3 billion people) and the Republic of India (with 1.21 billion). The third largest self-governing entity in the world, with a population of 600 million, was not a country but a social-networking site that had been created in two weeks by a young

Harvard student, Mark Zuckerberg, in 2004. It was originally called 'The Facebook' – after the practice common in many US universities of creating a reference book of students' names and photographs – but dropped the definite article in 2005 after purchasing the domain name Facebook.com for $200,000.[54]

Facebook's statistics are staggering. At the time of writing, half of its 600 million users were logging on at least once a day, and many were spending a significant amount of time on it. As well as posting 'status updates', browsing news feeds, photographs and videos posted by their friends and responding to invitations, they were uploading 200 million photographs a day, which explains why the site was hosting upwards of 90 billion photographs and was by far the biggest image-hosting site on the Internet.[55]

The most intriguing thing about Facebook, however, is not its phenomenal demographics, but why it has emerged as the dominant player in a long-established and crowded marketplace. The idea of computer-mediated social networking is, in fact, an old one which goes back to the Californian counterculture of the early 1970s; and the history of the field is littered with the remains of social-networking ventures that flourished spectacularly before either crashing to oblivion (SixDegrees) or finding a modest niche in the ecosystem (LinkedIn).

So what made Facebook special?

A brief history of social networking

The first thing to be said is that it wasn't the basic idea that made Facebook unique, because people had been exploring the use of computer networks for social networking from the time of the ARPAnet onwards. Social network(ing) sites (SNS) are defined by Danah Boyd and Nicole Ellison as 'web-based services that allow individuals to (1) construct a public or semi-public profile within a bounded system, (2) articulate a list of other users with whom they share a connection, and (3) view and traverse their list of connections and those made by others within the system.'[56]

The underlying idea was to use technology to connect people who share interests and activities across political, social, economic and geographical borders. A less lofty version was to create virtual spaces in which people could search for friends, relationships and business or employment opportunities.

The first SNS conforming to the Boyd/Ellison definition was SixDegrees.com which launched in 1997. It allowed users to create profiles, list their Friends and surf the Friends list. While it attracted millions of users, SixDegrees did not become a sustainable business and closed in 2000 just as the dot-com bubble burst. With hindsight, its failure might simply have been due to the fact that it was too far ahead of its time. 'While people were already flocking to the Internet,' write Boyd and Ellison, 'most did not have extended networks of friends who were online. Early adopters complained that there was little to do after accepting Friend requests, and most users were not interested in meeting strangers.'[57]

The years 1997–2000 saw a plethora of SNS-type services offering various combinations of profiles and publicly revealed friends. In 1999, a husband-and-wife team launched Friends Reunited in the UK. It was based on the idea of using the Internet to enable former school friends to make contact with one another, and effectively did for the UK what classmates.com had done in the US.

The next SNS wave began in 2001 when a business-oriented site, Ryze.com, launched. This was explicitly intended for people who wanted to expand their range of business contacts. In 2002 Friendster.com launched with the aim of competing in the market for dating services. Its distinctive offering was that while conventional dating services specialized in introducing strangers to one another, Friendster was designed to facilitate meetings between friends-of-friends (presumably on the assumption that such introductions are likely to prove more successful). The site attracted three groups of early adopters – bloggers, attendees at the Burning Man festival[58] and gay men – and grew very rapidly. But its server architecture proved unable to cope with the load imposed by this rapid expansion which – according to Boyd and Ellison – also led to 'a collapse in social contexts: users had to face their bosses and former classmates alongside their close friends'. The site continues in existence and claims to have 8.2 million members, but ninety per cent of its traffic now comes from Asia.

From 2003 onwards, SNS became mainstream, with a large number of profile-based sites appearing – among them LinkedIn, MyChurch, Xing, Orkut, Windows Live Spaces and MySpace. Of these MySpace appeared to be the most significant. Founded in 2003, it initially hoped to capitalize on Friendster's technical and other difficulties, but it rapidly became popular with teenagers because of its ingenious pursuit of connections with indie rock bands. By putting bands in touch with their fans, and vice versa, MySpace grew very rapidly and looked like becoming the dominant player in the SNS market.

This impression was confirmed when Rupert Murdoch,

proprietor of one of the world's most powerful media empires, bought the site for $580m in 2005. At the time, this appeared to many commentators to be an inspired move but in fact it proved to be a mistake, because MySpace had been put together by entrepreneurs who appeared to know relatively little about Web technology, and the technical limitations of the site (together with the fact that it allowed users to 'decorate' their profile pages with intensive graphics) meant that it often struggled to cope with the load imposed upon it by its users. But its biggest problem was that an unknown – and hitherto unregarded – competitor had appeared on the scene. And it came not from Silicon Valley, but from Harvard.

Most detached observers looking at MySpace and Facebook in early 2005 would not have seen them as direct competitors. For one thing, MySpace was much bigger in terms of web traffic. For another, it was riotous, vulgar and slightly weird. Because it allowed users to decorate their pages by adding all kinds of customized HTML code, MySpace pages were often a riot of animated gifs, flashing letters and colour-schemes that could only have been cooked up in workshops for the blind. Facebook, in contrast, seemed preppy, severe and rather dull. And it was exclusive – in that one had to belong to a smallish number of élite universities to be allowed to join. The contrast between the two sites led some observers to conclude that the social stratification of the real world had spread to cyberspace. One cartoon showed the preppy daughter of a wealthy New York banker family introducing her (scruffy) boyfriend to her parents. 'Don't you think he's a bit too MySpacey for you, dear?' says her mother.

And yet, in the end, it was MySpace that wilted and went

into decline. At the time of writing, its numbers are falling precipitously, despite corporate retrenchment, senior staff changes and at least one major relaunch.[59] It was the dull, preppy site from Harvard that emerged on top.

The Facebook story

In 2010 cinema audiences worldwide were captivated by a major Hollywood movie about a geek who became the world's youngest billionaire. It told the story of how a precocious Harvard undergraduate named Mark Zuckerberg took an idea essayed originally by three of his fellow students and turned it into a service that changed the world. The idea was social networking, and the service he created was Facebook.

The Social Network is a beautifully crafted, engrossing film written by Aaron Sorkin (author of *The West Wing*) and directed by David Fincher. It has some terrific performances – especially by Jesse Eisenberg who plays Zuckerberg as an obsessive, near-autistic genius reminiscent of the young Bill Gates. It was a big hit with audiences, received eight Oscar nominations (and won three) and garnered considerable critical acclaim.

The film was based on a best-selling book, *The Accidental Billionaires: Sex, Money, Betrayal and the Founding of Facebook*,[60] by Ben Mezrich. The story-line of the film is that, after a row with a girlfriend, Zuckerberg – then a sophomore specializing in computer science – storms back to his dorm room and in a frenzied burst of hacking, downloads images of female undergraduates and creates a website called Facemash which invites people to rank women by attractiveness. In a short time, the site has attracted 22,000 visits across the campus, generating abnormal traffic on the university's network. The Director of Computer Services is dragged out of bed early in the morning and shuts down the network. Zuckerberg is disciplined for violating university rules which state that 'Students may not attempt to circumvent security systems or to exploit or probe for security holes in any Harvard network or system, nor may students attempt any such activity against other systems accessed through Harvard's facilities' and for violating the university's guidelines for distribution of digitized images.[61]

As a result of the Facemash episode, Zuckerberg acquires a certain kind of celebrity status on campus. This brings him to the attention of two classic WASP[62] students, the Winklevoss twins, who, with a friend, Divya Narendra, have had an idea for a social-networking site for Harvard students. But as the two programmers they enlisted to write the code have left Harvard, they are stuck with unfinished software and are looking for someone to complete it.

They explain their needs to Zuckerberg, who signals his agreement to help, using the enigmatic expression 'I'm in'. But instead of writing code for the Winklevosses, Zuckerberg stalls them and creates his own social-networking site, which he calls 'The Facebook'. With the aid of some money supplied by a friend, Eduardo Saverin, he rents space on a commercial web hosting service and launches the service. The phenomenon that was to become Facebook is born.

Needless to say, the Winklevoss boys and Narendra are outraged, and eventually sue Zuckerberg for allegedly stealing their idea. His former friend, Saverin, also sues him for allowing the underhanded dilution of his (Saverin's) holding of Facebook stock. The story is told by interweaving the tale of

Facebook's evolution with scenes from the legal deposition hearings which eventually resulted in a settlement of the various disputes.

Real life rarely conforms to the dramatic requirements of screenplays, so it's not surprising that the film has been criticized for taking some liberties with the truth. The real-life Zuckerberg appears a more rounded, less autistic character than Eisenberg's portrayal of him; the row with the girlfriend that provides the dramatic opening to the movie is apparently entirely fictitious; it was not the Harvard network administrators who closed Facemash, but Zuckerberg himself; and so on. But broadly speaking its account of the genesis of Facebook seems fairly accurate.

Given that, it's an excellent example of the way the Internet enables 'permissionless innovation'. As the (Harvard) legal scholar, Lawrence Lessig, puts it, what's important about Zuckerberg's story is not that he's a boy genius:

> He plainly is, but many are. It's not that he's a socially clumsy (relative to the Harvard élite) boy genius. Every one of them is. And it's not that he invented an amazing product through hard work and insight that millions love. The history of American entrepreneurism is just that history, told with different technologies at different times and places. Instead, what's important here is that Zuckerberg's genius could be embraced by half-a-billion people within six years of its first being launched, without (and here is the critical bit) asking permission of anyone. The real story is not the invention. It is the platform that makes the invention sing. Zuckerberg didn't invent that platform. He was a hacker (a term of praise) who built for it.[63]

The 'hero' in this context is the network. Lessig points out that, for less than $1,000, Zuckerberg could get his idea onto the Internet. He needed no permission from the network provider, no clearance from Harvard to offer it to Harvard students, no permission from Yale to do the same with its students:

> Because the platform of the Internet is open and free, or in the language of the day, because it is a 'neutral network', a billion Mark Zuckerbergs have the opportunity to invent for the platform.[64]

But in a way, the permissionless innovation enabled by the Internet is only half of the Facebook story. Regardless of the rights and wrongs of the lawsuits and counter-suits filed by Zuckerberg's Harvard contemporaries, the fact is that the basic concept of online social networking was old-hat by 2004, when he began writing the code for the site. There's a yawning gap between a promising start-up and a $100 billion company[65] with 600 million users. So what remains to be explained is why Zuckerberg's venture succeeded where so many others faltered.

Looking back on it, one can see at least six factors at work:

- Unlike many other social-networking sites, Facebook started from a real space rather than from a virtual one. In this case it was a well-delineated community consisting of people whose email addresses ended with @harvard.edu and who for the most part lived and worked in close proximity to one another. The site expanded by incorporating other élite communities – initially from Ivy League universities and then from other higher education institutions, but in each case it was building on the fact that the incoming members belonged to real communities. In that sense, Facebook was building on the observation by Boyd and Ellison that 'what makes social network sites unique is not that they allow individuals to meet strangers, but rather that they enable users to articulate and make visible their social networks. This can result in connections between individuals that would not otherwise be made, but that is often not the goal, and these meetings are frequently between "latent ties" who share some offline connection.'[66]

- From the beginning, the technical implementation of Facebook seems to have been superb. This was initially probably a reflection of Zuckerberg's software skills, but it also suggests that the architecture of the system was sound from the outset. As a result, Facebook avoided the user alienation which plagued other sites (for example Friendster, MySpace and Twitter) that proved unable to scale up in the face of exponentially growing demand at a critical point in their development. And this remains true, in the sense that Facebook seems very robust while at the same time remaining a pretty lean operation in terms of engineering staff.
- Then there is the impact of Zuckerberg's personality on the evolution of his organization. The closest comparisons that come to mind are with Bill Gates and the co-founders of Google, Larry Page and Sergey Brin. One sees it in a transcendental ambition that goes beyond mere corporate objectives: in Gates's case it was 'a computer

on every desk, and each one running Microsoft software'; in the Google co-founders it is a desire 'to organize all the world's information'; with Zuckerberg it's a conviction that everything – including business – is social and that Facebook's destiny is to hold the world's **social graph**. Unlike many technically adept founders, however, (and in sharp contrast to the Google founders) Zuckerberg seems to have been willing from an early stage to hire professional managers, thereby freeing him to concentration on strategic issues.

- Allied to this has been a relentless focus on developing new services and features, some of which (e.g. news feeds from friends) have initially proved highly unpopular with users, plus a willingness to back off (as with some advertising initiatives) when it seemed prudent to do so.
- A key feature in the growth of Facebook was the decision to make it a software *platform* for which developers could write their own applications. This was akin to Bill Gates's insight that if Microsoft operating systems became the *de facto* standard for PCs, then a major ecosystem of software development would evolve around it, thereby locking customers into the Windows environment.
- And finally, there was the so-called *network effect* that is characteristic of many technological systems: the more people there were on Facebook, the more useful (or essential) it would seem to newcomers, and so a positive feedback loop kicks in leading to a 'winner takes all' outcome.

So . . . ?

These stories, in their different ways, provide an explanation of why the Internet has been such an enabler of innovation. In the first place, there's a talented individual with an idea that can be realized in software – which, after all, is pure 'thought stuff' – and the ability to write it. Secondly, there's a distribution system – the Internet – to which anyone can have access without having to seek permission or pay an entry fee. Thirdly, very little money is required in order to realize the idea and launch it on the world – often just enough to pay for web-hosting in the first instance. The result is an environment where the barriers to entry are incredibly low.

Many of the Internet ventures that have become household names (Amazon, eBay, Facebook, Flickr and Blogger, to name just five) were funded in their initial phase by their inventors' own resources and those of their families. Facebook's basic costs – until it had a quarter of a million users – were $85 a month for web hosting. (Mark Zuckerberg and his room-mate Dustin Moskovitz worked for no pay.) Jeff Bezos funded the first six months of Amazon.com's development out of his own savings. The founders of Flickr financed their start-up from personal funds, money from friends and family and a loan from the Canadian government.[67]

By freeing potential innovators from dependence on outside funding, the Internet's architecture, writes Barbara van Schewick,

> increases the number of innovative projects that might be realized. Many innovators will not be able to get outside

> funding, even if they want to, because they are not able to convince others of the viability of their idea. Many innovators whose applications later became very successful could not get venture capital or corporate funding when they first tried.[68]

The Web, Napster and malware were just three of the first-order surprises that have been sprung on us by Dr Cerf's Amazing Surprise-Generating Machine. There have been lots of others. For example Voice over IP (VoIP) – a fancy name for Internet telephony (the use of the network to carry voice calls) – and which has now become a mainstream technology in millions of homes across the planet. Then there's 'streaming media' – the use of the network to carry real-time audio and video (think YouTube; or the BBC iPlayer in the UK). Once upon a time, to listen to a radio station you had to live within a certain distance of its transmitters or to have sophisticated receiving and aerial technology. Nowadays I can listen over the Net to thousands of foreign stations, including Raidió na Gaeltachta, the Irish-language station based in Connemara on the west coast of Ireland, near where my father was born and many members of my extended family still live.

And on top of these first-order surprises we've seen a raft of others – like Wikipedia and Facebook – which are themselves the consequence of the open, permissive nature of the World Wide Web that Tim Berners-Lee built. In that way, the openness of the original architecture has effectively incubated other surprise-generating machines. And the chances are that it will go on like this for the foreseeable future.

The moral is that anyone who yearns for a period of stability – an intermission that will give us time to catch our collective breath and get a grip on things – is doomed to disappointment for as long as the original architecture of the system remains in place.

4. THINK ECOLOGY, NOT JUST ECONOMICS

Over the centuries, ever since the Victorian historian Thomas Carlyle christened it 'the dismal science', economics has had a bad press. The philosopher Bertrand Russell questioned its relevance when he caustically observed that 'economics is the study of how people make choices while sociology is the study of how they don't have any choices to make.' The banking crisis of 2007–8 darkened the discipline's reputation still further, because many people attributed the crisis to the naïve acceptance by banks, regulators and governments of economists' models of how risks should be evaluated and priced.[1]

Tools for thought

In dealing with anything complex, we need tools for thinking with – intellectual frameworks that enable us to represent and analyse the situation in abstract, conceptual terms. Traditionally, the framework of choice is provided by economics – that branch of social science that deals with the production and distribution and consumption of goods and services. It provides us with a way of thinking about supply, demand, markets and decision-making under conditions of uncertainty – although its assumptions about the human capacity for rational behaviour sometimes seem optimistic.

Nevertheless, economics is still the discipline that analysts reach for when they need to examine complex social systems in which millions of decentralized decisions determine outcomes. And in relation to our media environment, it has its uses – for example when regulators are looking at the concentration of ownership of traditional media organizations and the competitiveness (or otherwise) of markets. It's also relevant to managers of enterprises that produce information goods for profit: there are 'laws' of economics which cannot be suspended by wishful thinking – which is generally what happens in speculative bubbles.[2] But from our point of view, economics is deficient as an analytical framework in two important respects. Firstly, it is the study of the allocation of *scarce* resources, whereas the thing that most clearly distinguishes our emerging media environment is *abundance*, and traditional economics has little to say about that. Secondly, economic analysis assumes the primacy of markets, whereas much of the most significant activity in the emerging information environment appears to be taking place outside of the marketplace.

To illustrate this, let's look at the explosion of 'content' that has been triggered by the Internet, and particularly the Web. Once upon a time, if you wanted to be a publisher of content on anything other than a minuscule scale you had to own – or have access to – some pretty expensive (and therefore scarce) resources – for example: a newspaper, a printing press, a distribution network or a broadcasting station (and its associated licence from the government or official regulator). These realities led to the emergence of what the network scholar Yochai Benkler calls the 'industrial information economy' of the nineteenth and twentieth centuries[3] and its evolution followed the path that conventional economic analysis would have predicted: the emergence of competitive markets for paid-for content; increasing consolidation of ownership; regulatory intervention to ensure competition; and so on.

The arrival of the Internet and the personal computer has had a transformative impact on this industrialized economy. Suddenly, anyone with an Internet connection could become a global publisher by creating websites, writing blogs, publishing photographs, streaming audio or posting videos online.[4] The motto of YouTube is 'broadcast yourself'; and people do, on a vast scale – though often they are really re-broadcasting someone else's stuff.[5] A multitude of enthusiastic 'amateurs' could, in a few years, create the world's leading encyclopedia, Wikipedia. Another group – this time of skilled and dedicated programmers – could collaborate over the Internet to create one of the most sophisticated computer operating systems – Linux – a product far more complex than a modern airliner.[6] And so on. We are seeing, Benkler claims, the emergence of 'the networked information economy' in which 'decentralized individual action – specifically new and important cooperative and coordinated action carried out through radically distributed, nonmarket mechanisms that do not depend on proprietary strategies – plays a much greater role than it did, or could have, in the industrialized information economy'.[7]

Much of the content that is created by these processes of 'social production' (to use Benkler's phrase) is free, in the sense that it is *legitimately* available without charge.[8] A good deal of it is also free in the sense that it is published under licences that permit others to tinker with it and adapt it to their own uses. The significance of this is that here we have a burgeoning cornucopia of creative endeavours that are being undertaken outside of markets and apparently without the usual financial incentives that are the stuff of traditional economic analysis.

In a way, the intellectual frameworks provided by academic disciplines are like the colour filters that photographers use to absorb light of certain wavelengths and let only a portion of the visible part of the electromagnetic spectrum pass through. So you can have a filter that blocks everything except red light; or a yellow filter that enables photographers working in black and white to deepen the blue of the sky, thereby emphasizing and highlighting clouds. The point is that the scene being photographed remains the same, but different filters enable us to perceive it in different ways – to notice different aspects of it.

Economics is one such filter. It predisposes us to pay attention to certain things – markets, prices, competition, investment, returns and so on. But if there are no obvious markets, or activities going on for which there are no obvious financial returns, then economists tend to lose interest: their filter blocks what's happening.

The ecological imperative

An alternative filter derives from Marshall McLuhan, the Canadian cultural critic and aphorist who galvanized interest in media studies partly by borrowing a metaphor from science: the idea of media ecology, i.e. the study of media environments as ecosystems. In science, an ecosystem is 'a community of plants, animals and micro-organisms, along with their environment, that function together as a unit'.[9] In those terms, a *media ecosystem* can be seen as *a community of organizations, publishers, authors, end users and audiences, along with their environment, that function together as a unit.*

Looking at our media environment through an ecological filter leads us to a different kind of analysis. For example:

- It obliges us to treat it as a *system* distinguished by strong inter-relationships and dependencies between its constituent components. These components may look very different from one another but they are inextricably bound together. The system is, as the old adage puts it, 'greater than the sum of its parts'. This means that analyses based on so-called 'reductionist' studies of individual components taken in isolation are likely to be misleading. And change is *systemic*: when one component changes, the effects ripple through the entire system.
- It sensitizes us to the importance of *diversity*. In a natural ecosystem, different species take advantage of different niches which provide them with opportunities for specialized growth and success. And the complexity of an ecosystem ensures that there are countless niches for different roles and functions.
- It emphasizes the importance of *co-evolution*. Species migrate and adapt to fill available niches. Their adaptations lead to further change as one species changes the environment so that it becomes favourable for another. In the natural world, for example, lichens break down rock, which allows moss to grow, which in turn alters the environment so that grass can grow. In technology what happens is that one technology alters the environment (including attitudes and behaviour) so that other technologies become viable or acceptable.[10] Think, for example, of how the arrival of

domestic broadband connections made cloud computing a realistic proposition for non-corporate Internet users. As the anthropologist Bonnie Nardi puts it, 'A healthy ecology is not static, even when it is in equilibrium'. The media ecosystem is full of people who learn and adapt and create. 'The social and technical aspects of an environment co-evolve. People's activities and tools adjust and are adjusted in relation to each other, always attempting and never quite achieving a perfect fit. This is part of the dynamic balance achieved in healthy ecologies – a balance found in motion, not stillness.'[11]

- Finally, an ecological perspective alerts us to the importance of *keystone species* whose presence is critical to the survival of the ecosystem. In the case of a dune ecosystem, for example, marram grass – whose roots can go down twenty feet and are crucial in stabilizing the movement of sands in a windy environment – is a keystone species.

Case studies in media ecology

One of the defining characteristics of our media ecosystem is the explosive growth we have seen in 'user-generated content' – the publishing of blogs and home-made audio-visual material on the Web. The astonishing growth in blogging since 1990 prompted a predictable outburst of what John Seely Brown and Andrew Duguid call 'endism'[12] – as in questions about whether the phenomenon marks 'the end' of journalism. Yet, when one looks at it from an ecological perspective, one sees something much more interesting, namely co-evolution and the emergence of intriguing symbiotic relationships between old and new media. Here are some case studies that illustrate these emerging relationships.

Case study 1: Trent Lott and his big mouth.

On the afternoon of 5 December 2002, a group of Washington political insiders and their media acolytes gathered to celebrate the 100th birthday of the Senate's longest-serving member, Strom Thurmond. Those present included some Supreme Court justices, current and retired senators, political journalists, some members of President Bush's Cabinet – and the man who was due to be re-elected as Senate Majority Leader, Senator Trent Lott of Mississippi.

It was, by all accounts, a jovial occasion, with lots of joshing about Thurmond's advanced age and fabled lechery. There was, however, little mention of the old boy's racist past. In 1948 he and some of his southern brethren had split with Harry Truman over the latter's support for federal civil rights legislation. Thurmond had run for president as a 'Dixiecrat' candidate on a platform which espoused 'racial integrity' and segregation. 'All the laws of Washington,' he declared, 'and all the bayonets of the army cannot force the negro into our homes, our schools, our churches.'

But that was long ago, and Strom had allegedly mellowed with the passing years. Many people in the room were probably too young to remember the 1940s. Senator Lott, however, was not. He made a little speech in praise of Thurmond. 'When Strom Thurmond ran for president,' he declared, 'we voted for him. We're proud of it. And if the rest of the country had followed our lead, we wouldn't have had all these problems over all these years either.'

Although these remarks caused a certain *frisson* among the party-goers, they were not picked up in mainstream media reports of the occasion. The one exception was ABC News, which happened to have an observant young news reporter named Ed O'Keefe at the party. His organization ran a small piece about it on a company **blog**, but got no reaction. The story effectively died. Life inside the Washington Beltway went on as before, and in the normal course of events Senator Lott would have been re-elected Majority Leader.

Yet that wasn't what happened. On 20 December – a mere fifteen days after the party – Lott announced that he would step down as Majority Leader. The reason? A growing frenzy in mainstream media about his apparent endorsement of segregationism, and a stinging rebuke by President Bush delivered on 12 December in a speech to a largely black audience.

So what happened between 5 December and 20 December? The answer was that although the Lott story effectively 'died' in the mainstream media, it was kept alive in the **blogosphere** – the subculture of online diarists which has become the Net's version of Speaker's Corner – the area in London's Hyde Park where unrestricted public speaking has traditionally flourished. The realization of this led to an outbreak of triumphalism in the blogging community. It was portrayed as a demonstration of the superiority of online discussion *vis-à-vis* the narrowed agendas of old-style journalism. Traditional media advocates responded by pointing out that it was not bloggers' outrage but the crescendo of coverage in newspapers and broadcast media that motivated President Bush to make his speech throwing Lott to the wolves.

So who should get the credit for unhorsing the Senate Majority Leader?

A detailed study by a Harvard academic, Esther Scott, suggested that the truth was more complicated – and more interesting.[13] Her analysis of the controversy's chronology unearthed a more realistic picture of the relationship between mainstream journalism and blogging.

As Scott tells it, the mainstream media did indeed fail to pick up on the Lott story. There were various reasons for this. Most of the journalists at the party were 'insiders', long inured to the prejudices of politicians – so Lott's offhand remarks did not seem to the professional journalists present to be 'news'; and besides, the majority of them were probably too young to have known anything about racial politics in the American South in the immediate post-war years. But the main reason Lott's indiscretion effectively disappeared from public view was because traditional media need some way of keeping that kind of story alive. If they can't raise reactions from other public figures, then they have no justification (other than excessive editorial zeal, for which they may be criticized) for keeping a story going.

Bloggers, in contrast, labour under no such disadvantage: they can chew on a bone for as long as they like. It just happened that this particular bone was thrown to them by a traditional reporter. It turned out that some bloggers knew quite a lot about the history of racial segregation in the American South. As Jay Rosen, the eminent journalism professor put it:

> What news was heard that night when Lott rose to speak depended entirely on prior knowledge that a journalist either did or did not have about race, Southern politics, and the re-alignment of the political parties during the civil rights era. It also helped to know something of Trent Lott's career

> in Mississippi, going back to his college years at Ole Miss. If you didn't, you wouldn't hear anything special in his praise of Thurmond and the Dixiecrats of 1948. A party platform Edsall [a blogger] dug up summarizes it well: 'We stand for the segregation of the races and the racial integrity of each race.
>
> Well, bloggers did know this background, and they went right to it. Atrios, for example, printed the official Democratic Party ballot in Mississippi from 1948. 'Get in the fight for State's rights – fight for Thurmond and Wright,' it said.[14]

Or, as the Harvard study puts it, 'Bloggers weighed in quickly on Lott, offering readers a short course on Dixiecrat politics and their own acid commentary on the matter.'

But just as bloggers attend to traditional media, reporters read blogs, and it was the persistence of the story in the blogosphere that finally persuaded the big guns of US journalism – particularly newspapers like the *New York Times* and the *Washington Post* – to reopen it. After that, Lott's fate was effectively sealed. He resigned as Senate Republican Leader and left the Senate altogether in December 2007 to become a Washington-based political lobbyist.

Case Study 2: Dan Rather and the guys in pyjamas

On 8 September 2004, the American CBS television network screened a startling edition of its flagship documentary series *60 Minutes*. The programme was presented by the veteran TV reporter, Dan Rather. Its subject matter was the military record of the sitting President, George W. Bush – who was then running for re-election – during his time in the National Guard in the late 1960s and early 1970s.

The young Bush managed to avoid being drafted to serve in the Vietnam War by volunteering for duty in the Texas Air National Guard. He joined in May 1968 and committed to serve until May 1974, with two years on active duty while training to fly followed by four years on part-time duty. A central allegation of the *60 Minutes* documentary was that Bush's adherence to his military commitments had been erratic. The claims were based on copies of memoranda allegedly written by Bush's commander, Lt. Colonel Jerry Killian, which suggested that Bush had disobeyed orders during his national service, and that undue influence had been exerted on his behalf to 'improve' his record. There were, it was claimed, six of these documents, and *60 Minutes* claimed that four of them had been authenticated by experts retained by CBS.

In the context of a rancorous presidential campaign, these were sensational allegations – especially given that they concerned a president who had committed many thousands of young American lives to a controversial war in Iraq. It had long been claimed by Bush opponents that his influential family connections had enabled him to dodge the Vietnam draft by enlisting in the National Guard. Now it seemed that he had partially reneged on even that safe option.

Before the programme was transmitted on 8 September, the CBS documentary team had consulted four forensic document examiners, seeking their opinions as to the authenticity of the memoranda. The experts concentrated mainly on the question of whether the signatures in the documents were authentic. One raised questions about the 'th' superscript (as in '111th'), and two of them recommended that the production team should consult a well-known expert on typewriter fonts, but he did not return the producer's call until 11a.m. on the day of transmission, and in the end his opinion was not sought. On balance, the production team decided that the documents were genuine.

So the programme was transmitted and immediately caused a storm. *60 Minutes* posted the documents on its website to back up the story. This made them available to the blogosphere which – as you might expect – contained some people who *were* real experts in typography. They rapidly established firstly that the memos cited in the TV programme were most likely to have been produced by Microsoft Word – software which did not exist until the early 1980s, after the IBM PC had appeared on the market; and, secondly, that the 'th' superscript could not have been produced by any typewriter commonly in use in 1972–73, when the documents were supposedly written.

Initially, CBS briskly rejected the bloggers' criticisms. On 9 September it issued a statement saying that the memos had been 'thoroughly investigated by independent experts, and we are convinced of their authenticity'.[15] The *60 Minutes* report, the statement continued, 'was not based solely on recovered documents, but rather on a preponderance of evidence, including documents that were provided by unimpeachable sources.' And in a memorable aside, CBS Executive Vice-President Jonathan Klein launched a withering attack on the network's tormentors. 'You couldn't have a starker contrast,' he said, 'between the multiple layers of checks and balances [at *60 Minutes*] and a guy sitting in his living room in his pajamas writing'.

But it turned out that, in this case, the guys in the 'pyjamasphere' got it right. On 20 September, CBS retracted the story and launched an independent inquiry into the affair. It concluded that the documents were indeed forgeries. Following the inquiry, the network fired the producer of the offending programme and asked three other producers connected with the story to resign. The fracas also cast a shadow on Dan Rather's glittering journalistic career: he retired as the anchorman and Managing Editor of the CBS Evening News in 2005; his last broadcast was Wednesday, 9 March of that year.[16] And he's had to live with the fact that the concluding chapter of his career has become known as Rathergate, after the infamous superscript that lay at the root of the problem.

Case study 3: WikiLeaks and the powers that be

Up to 2009, few people had heard of WikiLeaks, an international non-profit organization that publishes on its website (www.wikileaks.ch) submissions of private, secret and classified media from anonymous news sources, news leaks and whistleblowers.[17] A year later the challenge would be to find people who hadn't heard of it. What made the difference was the fact that, until then, WikiLeaks had been publishing stuff that was mainly awkward for corporations and repressive political regimes in faraway countries. But in 2010 it began releasing material that was embarrassing to the United States, and that proved to be pivotal. The story of the evolution of WikiLeaks, and in particular of its changing relationship with traditional news media, provides another interesting case-study in media ecology.

WikiLeaks was founded by an Australian activist, Julian Assange, and others in October 2006 and released its first cache of documents at the end of that year. The early documents referred to events in Africa, including a copy of a decision by the rebel leader in Somalia to assassinate officials working for the Somali government.[18] In 2007, the site published documentation of corruption by the Kenyan leader Daniel Arap Moi.

The first WikiLeaks publications relating to the US came in November 2007 in the form of a copy of the 'Standard Operating Procedures' for the Guantanamo Bay detention site where terrorist suspects are interned, and an investigative report written by Julian Assange claiming that the US had 'almost certainly' violated the terms of the Chemical Weapons Convention as originally drafted by the UK in 1997. The article asserted that the deployment of CS munitions and dispensing equipment and weapons capable of firing CS gas by the US was a violation of the Convention.

In 2008, WikiLeaks published: details of a Swiss Bank's tax avoidance activities in the Cayman Islands; internal documents of the pathologically secretive Church of Scientology; a copy of the Apple iPhone application developer contract (a document which developers are legally barred from discussing in public); US military Rules of Engagement in Iraq permitting cross-border pursuit into Iran and Syria of former members of the Saddam Hussein regime; an early draft of the Anti-Counterfeiting Trade Agreement; emails from Sarah Palin's Yahoo accounts when she was running for vice-president; a membership list of the British National Party; and documents about extra-judicial killings in Kenya, for which WikiLeaks was awarded Amnesty International's New Media Award.

In 2009, WikiLeaks was even more active, releasing material documenting corruption, wrongdoing and incompetence in a range of countries and organizations, including oil-industry corruption in Peru; banking abuses in Iceland; a nuclear accident in Iran; and a trove of emails from climate scientists at the University of East Anglia in the UK.

Summarizing these activities, the American legal scholar Yochai Benkler observed that, prior to 2010, 'Wikileaks was an organisation that seems to have functioned very much as it described itself: a place where documents that shed light on powerful governments or corporations anywhere in the world, or, in the case of the climate scientists' emails, on a matter of enormous public concern, could be aired publicly'.[19]

The year 2010 was a pivotal year for the site. In March it published a secret US Department of Defense report, written in 2008, which identified WikiLeaks as a danger while at the same time recognizing it as a source of investigative reporting about US military procurement and behaviour. This was followed from April onwards by four separate disclosures of documents and materials, all apparently based on a trove of materials provided by a 22-year-old US Army private, Bradley Manning. The first release was a video entitled 'Collateral Murder' based on a gunsight video-recording made by a US Apache helicopter gunship during an incident on 12 July 2007. The Apache fired on a group of individuals in Iraq, killing about twelve, two of whom happened to be employees of the Reuters news agency.

The *Collateral Murder* video marked a new departure for WikiLeaks in two ways: it was an intensively edited production, released at a conventional news conference held at the National Press Club in the US capital; and the way the film had been edited suggested a possible shift into 'advocacy journalism'. The furore over the latter aspect of the film was tacitly acknowledged by the subsequent release of the (uncut) original footage of the incident.

The second publication of sensitive US material came in July 2010, when WikiLeaks released so-called 'war logs' from US forces in the field in Afghanistan. For this, the site adopted a radically different mode of presentation. Instead of just putting the material on its website, this time WikiLeaks teamed up with three traditional news organizations – the London *Guardian*, the *New York Times* and Germany's *Der Spiegel*. These 'partner' organizations were given time to analyse, verify and prepare the material for publication. All four organizations then published on the same day, with WikiLeaks releasing the full cache of documents and the others publishing their analyses of them.

In October, WikiLeaks released a further cache of war-logs – in this case 440,000 field reports from Iraq which, among other things, revealed that Iraqi civilian casualties were higher than generally admitted, and that US forces, despite knowing that Iraq's military and police were systematically torturing prisoners, were making no systematic attempt to stop the practice. As with the Afghanistan war-logs, the release was done in partnership with the same traditional media organizations.

The final document release of 2010 came on 28 November and involved a limited and carefully sifted drip-feed from a trove of 250,000 confidential US diplomatic cables which was included in the material allegedly downloaded by Private Manning. This time the number of traditional news organizations involved was increased initially to four – the *Guardian*, *Der Spiegel*, *Le Monde* (France) and *El Pais* (Spain). The *Guardian* then made the documents available to the *New York Times*.

The interesting thing about this final release – which has since become known, inevitably, as 'Cablegate' – was the intensity and ferocity of the US reaction. Despite the absence of convincing evidence that the publication of the cables had inflicted significant damage on US interests or endangered the lives of individuals, WikiLeaks was subjected to hysterical criticism by some prominent American politicians. There were calls from some right-wing media figures for the assassination of Julian Assange. WikiLeaks was also subjected to sustained cyber-attack, was denied **DNS** services (which meant that the IP addresses of its sites could not be ascertained by anyone seeking them via a domain name), was denied hosting services by Amazon.com (one of the few Internet organizations with an infrastructure capable of withstanding heavy cyber-attacks), and had its conduits for accepting donations via PayPal and credit-card companies withdrawn. It was also reported that the US Justice Department was preparing the grounds for the extradition of Julian Assange – an Australian citizen – to the US to stand trial under the Espionage Act of 1917.

Benkler's analysis of the WikiLeaks story is extensive and covers the constitutional position (e.g. whether WikiLeaks would be entitled to First Amendment protection); the multi-faceted nature of the official and unofficial responses to the 'Cablegate' revelations; legal aspects of the story; and an analysis of the attitudes of the traditional media (the so-called Fourth Estate) towards WikiLeaks. For anyone interested in media ecology, however, two particular aspects of the story stand out.

An evolving symbiosis

The first concerns the way WikiLeaks's strategy has evolved over the years since its foundation. It began, remember, simply as a secure electronic 'dropbox' for confidential information, and an uncensorable site for publishing material supplied to it by whistle-blowers. So for a time it was a standalone entity making use of the disruptive potential of the Web for global dissemination of information and data. Given the obsessive secretiveness of many governments and corporations, the WikiLeaks founders may have originally assumed that the mere publication of such explosive material would be sufficient to achieve their desire 'to bring important news and information to the public' and 'to publish original source material alongside our news stories so readers and historians alike can see evidence of the truth'.[20]

In some cases, these hopes were realized. A few of its revelations (e.g. the Kenyan documents published in 2007 and 2008) did indeed have disruptive effects in the societies to which they referred. But much of the material published on the site appeared to have disappointingly little impact. It looked as though merely publishing masses of confidential information on a website was not a sure-fire way of gaining the world's attention. Facts, it seemed, did not always speak for themselves.

This seems to have prompted a quest within the WikiLeaks organization for a more effective way of publishing. In that context, the release, in 2010, of the *Collateral Murder* video can be seen as an experimental move towards a more traditional kind of advocacy journalism. But the experiment seems to have been regarded as a failure, and so the quest for a different publication model resumed.

The model that was eventually selected was that of 'partnering' with traditional news organizations – components of the Fourth Estate – which would bring to bear the skills of professional journalism (e.g. sifting, verifying, corroborating, redacting information that might endanger lives, etc.) on the material before it was released into the public domain.

The new 'partnership' model is interesting in ecological terms, because it's clearly symbiotic. That is to say, it's a relationship between two organisms which is of benefit to both. We can see this clearly by asking what each gets out of it.

For WikiLeaks, the relationship delivered several significant benefits:

- It ensured wider public attention for the war-logs and diplomatic cables by accessing the readerships – on both sides of the Atlantic – of well-known and influential traditional media organizations, and harnessing their institutional credibility.
- It ensured that the story would have 'legs', as journalists say – i.e. that it would not just be a one- or two-day sensation, because the traditional partners had an interest in keeping it in the public eye.
- It provided a kind of insurance that the powers that be would not succeed in suppressing the revelations. This was because the relationship leveraged the privileged position of the partners in their legal jurisdictions. In the US, for example, the *New York Times* would enjoy the protection of the First Amendment. In the UK, while legal protections for press freedom are weaker, it would still be difficult to suppress the *Guardian* without a protracted legal battle. Much the same applied to *Der Spiegel* in Germany (and, later, to *Le Monde* in France and *El Pais* in Spain).

For the established media partners, the relationship also offered significant benefits.

- It demonstrated their commitment to public-interest journalism by ensuring that a trove of important documents reached public attention.

- It provided a showcase for their professional skills in sifting, evaluating and validating controversial evidence – and in redacting information that might put individual lives at risk.
- It established their reporting credentials in a new public space and motivated them to begin thinking about how they might tool up for this new activity – e.g. by developing their own secure electronic dropboxes.
- And last, but by no means least, in a highly competitive market for public attention, it gave them a slice of some of the biggest journalistic action of the decade.

But in addition to the advantages for the partners in the relationship, there was also a system-wide benefit because, in a way, the partnership was the only way of dealing responsibly with the material. In cases like this, where there is the possibility that the lives of named or identifiable individuals could be endangered by indiscriminate disclosure, powerful networked methodologies like crowdsourcing simply won't work. As Benkler puts it:

> Looking specifically at Wikileaks and the embassy cables shows that responsible disclosure was the problem created by these documents that was uniquely difficult to solve in an open networked model. That problem was not how to release them indiscriminately. That is trivial to do in a network. The problem was how to construct a system for sifting through these documents and identifying useful insights. Protestations from the professional press that simply sifting through thousands of documents and identifying interesting stories cannot be done by amateurs sound largely like protestations from *Britannica* editors that Wikipedia will never be an acceptable substitute

for *Britannica*. At this stage of our understanding of the networked information economy, we know full well that distributed solutions can solve complex information production problems. It was the decision to preserve confidentiality that made the usual approach to achieving large-scale tasks in the networked environment – peer production, large-scale distributed collaboration – unavailable. One cannot harness thousands of volunteers on an open networked platform to identify what information needs to be kept secret. To get around that problem, Wikileaks needed the partnership with major players in the incumbent media system, however rocky and difficult to sustain it turned out to be.[21]

Hostility of the incumbents

One of the most noticeable aspects of Cablegate was the fact that many – if not most – traditional media organizations were either unsupportive of the venture (though content to relay its revelations to their own audiences) or downright hostile to WikiLeaks. Hostility was evident in a number of ways. Firstly, there was the consistent and widespread repetition of the assertion that WikiLeaks had irresponsibly 'dumped' 250,000 documents in the public domain with no thought to the consequences for individuals whose lives might be endangered in the process. Benkler finds that 68 out of 111 stories covering the release of the embassy cables in major US newspapers made these kinds of claims, and a further 20 were 'ambiguous'.[22] The claims were palpably untrue: only a small fraction of the 250,000 cables reached the public domain through the filter created by the partnering organizations. It's worth noting that in 2011, after Assange's relationships with his newspaper 'partners' disintegrated, the entire trove of documents was released in unredacted form. It's not clear how or why this happened.[23]

More significant was the way in which traditional media connived in governmental attempts to situate WikiLeaks in a politically convenient context. The defining features of these efforts, writes Benkler, was to frame the event, not as journalism, but as 'a dangerous, anarchic attack on the model of the super-empowered networks of terrorism out to attack the US'. He cites Hilary Clinton's statement of 30 November 2010. 'Let's be clear,' said the Secretary of State. 'This disclosure is not just an attack on America's foreign interests. It is an attack on the international community – the alliances and partnerships, the conversations and negotiations, that safeguard global security and advance economic prosperity.'[24] The US Vice-President, Joe Biden, echoed this line when he described the WikiLeaks founder, Julian Assange, as 'a high-tech terrorist'.[25]

Much of the traditional media picked up the message. The *New York Times* columnist Thomas Friedman, for example, embarked on a peculiarly irrational rant on 14 December 2010. 'The world system,' he wrote,

> is currently being challenged by two new forces: a rising superpower, called China, and a rising collection of superempowered individuals, as represented by the WikiLeakers, among others. What globalization, technological integration and the general flattening of the world have done is to superempower individuals to such a degree that they can actually challenge any hierarchy – from a global bank to a nation state – as individuals.[26]

Later in the same piece there is this extraordinary passage: 'As for the superempowered individuals – some are constructive, some are destructive. I read many WikiLeaks and learned some useful things. But their release also raises some troubling questions. I don't want to live in a country where they throw whistle-blowers in jail. That's China.'

Among other things, Friedman seems oblivious to the fact that Bradley Manning, the alleged whistleblower in the war-logs and Cablegate cases, was being held in the US under conditions that most countries and many lawyers would regard as torture.[27]

The hidden subtext of much of the traditional media commentary on WikiLeaks was: 'Wikileaks is not journalism, but part of a wider conspiracy against the US.' The message was further reinforced by hostile reporting of Julian Assange, WikiLeaks's charismatic and enigmatic founder. Here for example, is the opening passage of the profile published in the *New York Times* in October 2010:

> Julian Assange moves like a hunted man. In a noisy Ethiopian restaurant in London's rundown Paddington district, he pitches his voice barely above a whisper to foil the Western intelligence agencies he fears.
>
> He demands that his dwindling number of loyalists use expensive encrypted cellphones and swaps his own the way other men change shirts. He checks into hotels under false names, dyes his hair, sleeps on sofas and floors, and uses cash instead of credit cards, often borrowed from friends.[28]

Here we have the essential ingredients of an emerging media 'narrative': the WikiLeaks founder as a shifty, untrustworthy person. This narrative was further embroidered by the *New York Times* Executive Editor, Bill Keller, in an 8,000-word magazine essay that later served as the foreword to an ebook issued by the paper about the cables. Keller introduces Assange as 'an eccentric former computer hacker of Australian birth and no fixed abode' and then recapitulates the impressions of another NYT journalist who reported that 'Assange slouched into the *Guardian*'s office a day late . . . He was alert but dishevelled, like a bag lady walking in off the street, wearing a dingy, light-colored sport coat and cargo pants, dirty white shirt, beat-up sneakers and filthy white socks that collapsed around his ankles. He smelled as if he hadn't bathed in days.'[29]

What's interesting about the tone of this stuff is not so much the palpable animosity towards Assange, who is clearly a complex character with whom everyone involved had difficulties from time to time, but that it comes from a senior executive of a traditional journalistic organization that was a *partner* in the publication of the embassy cables.

A closer reading of Keller's diatribe, however, suggests a revealing subliminal message – a desire to put clear blue water between WikiLeaks and the *New York Times*. Keller states that one of his prime concerns was 'how we would ensure an appropriate distance from Julian Assange. We regarded Assange throughout as a source, not as a partner or collaborator, but he was a man who clearly had his own agenda'. At another point, Keller says: 'I do not regard Assange as a partner, and I would hesitate to describe what Wikileaks does as journalism.'[30] This is translated by Benkler as asserting a categorical distinction between the *New York Times* as an organizational form and the decentralized, networked form represented by Wikileaks. In fact Keller's essay is noticeable for its implicit determination to emphasize the dichotomy between professional/reliable journalism (the *New York Times*, *Guardian*, *Der Spiegel*) on the one hand, and unprofessional/unreliable WikiLeaks on the other.

To a media ecologist what this highlights, of course, is how difficult the dominant media organisms of the old ecosystem are finding it to cope with this new, networked intruder. And although it's clear at one level that what has to happen is the evolution of new, symbiotic relationships between networked and more traditional kinds of journalism, the antagonism towards Assange suggests that, at a visceral level, the incumbents see it as a **zero-sum game**. At any rate, by conniving in the framing of WikiLeaks as part of the terrorist threat rather than as a new form of journalism, they were in effect making the newcomer more vulnerable to the multifaceted and extra-legal threats that it faced – and less able to shelter behind the constitutional protections that the incumbents themselves enjoy.

News as an ongoing conversation

These case studies illustrate one important aspect of what's happening to our media ecosystem. A new species has arrived, and existing species are having to accommodate themselves to the newcomer. And *vice versa*. The case studies do not support the claim that so-called 'new' media are wiping out their older counterparts. The blogosphere did not bring down Trent Lott; that was achieved by the power of 'old' media, particularly traditional newspapers and broadcast media, to shape the public agenda – to create a situation in which the Senate Majority Leader was effectively deemed untouchable by existing political structures. But traditional media would probably not have been able to create that situation had not the story been kept alive by the blogosphere. So what we're seeing is the emergence of what an ecologist would see as a symbiotic relationship.

The lessons of Rathergate are different. It's conceivable that traditional media – in the form of major newspapers or other broadcasting organizations – would have been able to refute the allegations of the *60 Minutes* broadcast by locating and deploying experts in typography who could convincingly establish that the Bush memoranda were forgeries. What was striking is the speed with which the blogosphere was able to achieve this result by harnessing the collective intelligence and expertise of the network. The prominent blogger Ken Lane summarized this succinctly when he declared 'We can fact-check your ass'.[31] In that sense, the story illustrates the extent to which the blogosphere has become a check on the power of conventional media.

The story also illustrates the dangers for traditional media of ignoring the online world – of underestimating the collective intelligence of its audiences. As Jay Rosen puts it:

> Another big reason why *60 Minutes* wound up with Rathergate was that [CBS News] didn't have the Web literacy that they should have had. . . . They didn't know what was happening to their story on the Web. They went into a defensive crouch and said, 'We stand by our story.' But nobody in power, nobody with any decision-making influence actually was monitoring what was happening to that story online, and they didn't know how weak their story was. They persisted for at least seven or eight days in something that many other people knew was going to fall . . . It was a case of a news organization that was not actually current with knowledge of its own story. It was extraordinary.[32]

But what all three case studies unambiguously demonstrate is the extent to which the media world has changed. And this is just part of a much bigger story.

Our evolving media ecosystem

If you start to look at our media environment in this way then the first thing you ask is: what are the species in this system? And how do they co-exist with one another? Until Gutenberg, the media ecosystem was pretty simple: most communication was oral, and public communications were limited to what could be achieved by addressing gatherings of people, mainly in churches, temples or stadia. The advent of printing increased the complexity of the system by adding a range of new 'species': publishers of, first, books, and then flyers, leaflets, pamphlets and ultimately magazines and newspapers. The electric telegraph added another layer of complexity by collapsing distance and ratcheting up the speed at which information circulated round the system.[33] Then came the telephone and – more significantly – radio, the first truly *broadcast*[34] medium.

Although it was later to be overshadowed by television, radio had a profound effect on our media ecosystem. When it first appeared in the US in the 1920s, nobody had a business model for it: broadcast signals, by definition, went out into the ether, and anyone who had the appropriate receiving equipment could listen to them, for free. How could one make money from that? The first answer was to manufacture and sell radio receivers, and initially that business thrived but eventually tailed off as the market became saturated. Hundreds of companies were founded – and went bust – in a frantic collective attempt to find a sustainable business model for the new medium. In the end, after rather more than a decade, the problem was solved – by a soap manufacturer, Procter and Gamble. Its executives figured out that if one sponsored the production of compelling dramatic material for radio, then commercial advantage would follow from being associated with this in the public mind. Thus was born the *soap opera* – a serialized drama transmitted in daily or weekly intervals and featuring continuing story-lines. The first such production – a series called *Guiding Light* – was broadcast in January 1937.

Radio created the world of mass media because it enabled broadcasters (and advertisers) to reach mass audiences for the first time. It effectively welded the United States, hitherto a collection of regional markets spread over several time zones, into one unified, continental market. And it enabled audiences of many millions of geographically dispersed listeners to share the same listening experience. It was the birth of what later became known as 'water-cooler media'.

One of the laws of communications technology is that new media are generally additive rather than substitutive, which is a fancy way of saying that new technologies generally don't wipe out older ones. That's not to say that sometimes media don't disappear. Printing *did* effectively end the scribal tradition, and encyclopedias on CD-ROM (and, later, on the Web) effectively ended the dominance of printed reference works. But these are exceptions that prove the rule. In general, the sequence goes more like this: new medium appears in the ecosystem; this challenges the position of existing media, which then either adapt to the newcomer or fade into irrelevance; and, over time, the ecosystem reaches a new equilibrium state.

What this means is that 'endism' is misleading when studying the evolution of media ecosystems. Television did not kill off radio – or movies, for that matter. Electronic encyclopedias killed off printed works of reference, but left the novel untouched. E-readers like the Amazon Kindle or the Apple iPad will doubtless have an impact on the market for printed books, but they are unlikely to wipe them out. And so on. In each case the story is the same: new media don't wipe out old media. But their arrival does change the ecosystem.

A good case study is provided by the arrival of broadcast television in Britain in the 1950s. At that time, most citizens got their news either from radio or from newspapers. But as ownership of TV sets grew, and a commercial channel emerged to challenge the BBC monopoly, an increasing number of people looked to the new medium for news. This posed a direct challenge to newspapers: if they were no longer to provide news, what could they offer? They responded in one of two ways. The popular – mass-market – newspapers adapted to the new medium by becoming parasitic on it – feeding on the cult of instant celebrity spawned by it, much as they had with Hollywood movies in the 1930s and 1940s. The so-called 'quality' papers (the ones we used to call broadsheets) responded in quite a different way: by offering increasing amounts of commentary and interpretation rather than raw news[35] and, later, 'feature' articles. TV didn't kill newspapers, in other words; but its arrival in the ecosystem forced them to change.

As it happens, the arrival of the Internet, and particularly the Web, has proved a much bigger threat to print newspapers than television ever was. This is because it has dramatically undermined their business model. Newspapers – like most corporate enterprises – are *value chains*.[36] That is to say, they link unprofitable activities (cost centres) with profitable ones in such a way that revenues from the latter exceed the costs of the former. In the case of newspapers, journalism – the gathering, collating, editing, publishing of news and opinion – is an expensive, loss-making activity. Advertising – display and classified – on the other hand was highly profitable. But in order to attain the circulations that would make advertising profitable, one needed all that expensive, loss-making journalism: it was what attracted readers and sold newspapers, providing sales revenues and notching up the rates one could charge advertisers.

When the Web first appeared, newspapers thought that the main threat it posed would come from online *news* distributed for free. They were wrong: it turned out that the bigger threat was to the profitable part of their value-chain – classified advertisements. This is because what the Internet does is *dissolve* value chains. The first signs of this came in 1995, when an entrepreneur named Craig Newmark started an email list featuring events in San Francisco. A year later, the email service morphed into a website, Craigslist.com, which allowed people to post free classified ads while charging for job and property listings. From its original base in San Francisco, Craigslist steadily expanded to cater for an increasing number of cities, initially in the continental United States, but later in many other countries.[37]

Wherever it opened a site for a new location, the impact of Craigslist on newspaper revenues was dramatic. By 2004, for example, it was reckoned that it was siphoning off between $50m to $65m in advertising revenues from print publications in the San Francisco Bay area alone. 'In the Bay area,' wrote Bob Cauthorn, a former media executive at the *San Francisco Chronicle*,

> if you're looking for a job, a house, an apartment; anything to put in that house or apartment; basically, if you want anything a classified market can provide, you don't need to go anywhere

but Craigslist. Indeed, if you want to be sure you've seen it all, you must use Craigslist. And you can pretty safely ignore the print newspapers.[38]

Of course the success of Craigslist was partly due to the fact that personal listings were free – as was access to the site. But it was more to do with the fact that classified advertising simply works much better on the Web. Instead of having to wade through pages of densely printed ads in 9-point type looking for that used car or that chic studio apartment in a certain neighbourhood, all you have to do is type a query in a search box and – bingo! – there's a list of possibilities. Given that, the surprising thing about the Web is not that it so radically undermined the economics of print newspapers, but that they didn't see it coming. They failed to perceive the real threat to their value chain.

Up to 1990 our media ecosystem was largely dominated by broadcast (few-to-many) media, and especially by television. TV was the medium that absorbed most public attention, shaped political discourse and dominated entertainment and sport.[39] And although print media – or at least the segment of them represented by newspapers and magazines – also had an important impact on public life and culture, they shared with broadcast one central, defining feature – namely that they were also centrally owned and edited. In that sense, the dominant media functioned as cultural gatekeepers. This also applied to the film industry. What this meant is that in the ecosystem that had endured from the 1920s up to the 1990s, nobody could get published widely unless s/he had been approved by these gatekeepers. Or, as one cynic (I think it was A.J. Liebling) put it, freedom of the press was available to anyone who was rich enough to own a newspaper.

The other noticeable feature of the 1990 ecosystem is the relatively minor role played by mobile phones and the Internet. Remember that in 1990, the Web was still just a project of Tim Berners-Lee at CERN. It didn't begin to go mainstream until the spring of 1993.

In the second decade of the twenty-first century, our media ecosystem looks very different. Its most significant aspects are: a huge increase in the relative importance of the Internet, which has grown to overshadow that of broadcast media; a relative decline in the importance of traditional print media; and a much more dominant role for mobile phones.

My hunch is that over the next decade, the Internet will move to become the dominant 'species' in our ecosystem. That doesn't mean that other, older media will wither away, just that their places in the ecosystem will change as they adjust to the new reality. So the interesting question becomes: what will a world dominated by the Internet be like?

Properties of a Net-centric ecosystem

In thinking about this, two useful words are 'push' and 'pull' because they capture the essence of where we've been and where we seem to be headed.

Broadcast TV is a 'push' medium: a relatively select band of producers (broadcasters) decide what content is to be created, create it and then *push* it down analogue or digital channels at audiences which are assumed to consist of essentially passive recipients.

The couch potato was, *par excellence*, a creature of this world. He did, of course, have some freedom of action. He could choose to switch off the TV, but if he decided to leave it on, then essentially his freedom of action was confined to choosing from a menu of options decided for him by others, and to 'consuming' their content at times decided by them. He was, in other words, a human surrogate for one of the behaviourist B.F. Skinner's pigeons – free to peck at whatever coloured lever took his fancy, but not free at all in comparison with his fellow pigeon perched outside on the roof.[40]

Another distinctive feature of the world of push media was its fundamental asymmetry. All the creative energy was assumed to be located at one end (the producer/broadcaster). The viewer/listener was assumed to be incapable of, or uninterested in, creating content; and even if it turned out that s/he was capable of creative activity, there was no way in which anything s/he produced could have been published.

Looking back, the most astonishing thing about the broadcast-dominated world was how successful it was for so long in keeping billions of people in thrall. Networks could regularly pull in mass audiences (in the tens of millions) for popular programmes – and pitch their advertising rates accordingly. Small wonder that one owner of a UK ITV franchise (Roy Thompson) famously described (in public) commercial television as 'a licence to print money'.

But in fact the dominance of the push model was an artifact of the state of technology. Analogue transmission systems severely limited the number of channels that could be broadcast through the ether, so consumer choice was restricted by the laws of analogue electronics.[41] The advent of (analogue) cable and satellite transmission and, later, digital technology changed all that and began to hollow-out the broadcast model from within.

The Web is the opposite of broadcast: it's a *pull* medium.

You choose stuff and click on it to pull it down onto your computer. You're in charge. In the words of a leading broadcast executive, the Web is a 'lean forward' medium, in contrast to TV which is a 'lean back' medium.

So the first implication of the switch from push to pull is a growth in consumer sovereignty. We saw this early on in e-commerce, because it became easy to compare online prices and locate the most competitive suppliers from the comfort of your own armchair. But the Internet doesn't just enable people to become more fickle and choosy consumers. It also makes them much better informed – or at least provides them with formidable resources with which to become more knowledgeable.

The Net is also making it much harder for organizations and companies to keep secrets. If one of your products has flaws, or if a service you provide is sub-standard, then the chances are that the news will appear somewhere on a blog or a posting to a newsgroup or email list. And when it does, conventional PR news-management techniques are ineffective – as many companies now know to their cost.[42]

Creativity 2.0

The emergence of a truly sovereign, informed consumer is thus one of the implications of an Internet-centric world. This is significant, of course, but it was entirely predictable, given the nature of the technology. And in the end it may turn out to be the least interesting part of the story – because another significant consequence of an Internet-centric world lies not in the arena of consumption, but in production. In blunt terms, the asymmetry of the old, push-media-dominated ecosystem is being replaced by something rather more balanced and multi-way.

Remember that the implicit assumption of the broadcast model was that audiences are passive and uncreative. As the Web went mainstream, we discovered that this passivity and apparent lack of creativity might have been more due to the absence of tools and publication opportunities than to intrinsic defects in human nature.

Take blogging – the practice of keeping an online diary. There are currently millions of blogs – personal, corporate and collaborative. Many of them are examples of vanity publishing with little literary or intellectual merit. But hundreds of thousands of blogs are updated daily – sometimes hourly – and many contain writing and thinking of a very high order. In my own areas of professional interest, for example, blogs are some of my most trusted sources, because I know the people who write them, and some of them are leading experts in their fields or interesting and original thinkers.

What is significant about the blogging phenomenon is its demonstration that the traffic in ideas and cultural products isn't a one-way street, as it was in the old push-media ecosystem. People have always been thoughtful, articulate and well-informed, but up to now relatively few of them ever made it past the gatekeepers who controlled access to publication media. Blogging software and the Internet gave them the platform they needed – and they have grasped the opportunity in very large numbers.

A similar pattern has been evident with user-generated images and videos. At the time of writing, the leading image-hosting site Flickr.com hosts over five billion photographs,[43] and is growing at a rate of several thousand images an hour. The vast majority of these are photographs taken by 'amateurs', and a significant proportion are published under Creative Commons licences which enable varying degrees of (generally non-commercial) re-use.[44]

We've seen much the same thing with user-generated video and the rise of video-hosting sites like YouTube. The scale of the YouTube phenomenon is staggering: an average of 48 hours'-worth of video is uploaded to the service every minute. This means that it also hosts an astonishing range of content, from the ridiculous to the sublime and everything in between.

From the outset, YouTube has been the target of litigation by movie and TV companies and other branches of the multi-media 'content' industries, on the grounds that many of the video clips uploaded to the site infringe copyright.[45] There is undoubtedly a lot of infringing content on the site (though it's difficult to get an accurate estimate of the relative proportion of the total that violates copyright), but the mainstream media's obsession with so-called 'theft' obscures the fact that much of the stuff on YouTube is original and non-infringing. A classic case in point is 'Charlie Bit My Finger – Again!',[46] a charming 56-second video uploaded to YouTube in May 2007, and which in three years has been viewed over 190 million times. It shows a young boy and his baby brother and was uploaded by their father so that it could be viewed by their godfather. There was simply no way in the old ecosystem that a simple clip from a home movie like this could have garnered such worldwide interest.

Another phenomenon that has been sidelined by media obsession with intellectual property is the growth of 'remix culture'[47] – as evidenced, for example, in what has been enabled by the combination of inexpensive video-recording/editing technology with the publication opportunities provided by YouTube.

A celebrated case in point is what has been done with a particular clip from the film *Downfall*, a feature film that charts the last days of Adolf Hitler in his Berlin bunker.[48] The pivotal moment in the film comes when the dictator's senior generals reveal to him that the game's up – that the Allies have Berlin encircled, that the Russians are at the city's eastern gates and that German military resources are exhausted.

In the scene, Hitler (played by Bruno Ganz) absorbs this information in silence. Then he orders most of those present to leave the room, leaving only the most senior military people. He looks crushed, broken. And then he explodes into life. As David Denby, the *New Yorker*'s film critic, describes it,

> Energy rushes up in galvanic surges from Ganz's pelvis or spine or some other mysterious source of actorly strength. The dark head, slumped over a map, suddenly rises, the arms wave about wildly, and the voice erupts in that familiar deafening bawl. The rages are mesmerizing, appalling . . . and somehow Ganz pulls them off without lunging all over the room; he explodes and implodes simultaneously, and then subsides and becomes even smaller.[49]

Ganz's performance is a real *tour de force* – so much so that the *New Yorker* critic wondered aloud if it would have the effect of humanizing Hitler. It gives one the feeling that *this is what the monster must have been really like*. But the scene had another, equally extraordinary, side-effect. It became the basis for a wildly successful and entertaining comic virus, in which people used everyday video-editing software to remix the scene in modern contexts (politics, sports, technology, popular culture). The German soundtrack was left unchanged, but new subtitles were added, and then the results were posted on YouTube.

So we had Hitler raving about the defeat of the New York Mets, or being excluded from XBox Live or not getting an Apple iPad. After Irish voters rejected the Lisbon Treaty in a referendum, Hitler became the Irish *Taoiseach* [Prime Minister], Brian Cowen, ranting about the venality and incompetence of his advisers and the stupidity of the electorate. In one spoof, Hitler becomes Hillary Clinton who is enraged by Barack Obama's victories in the US presidential primaries; in another, he is John McCain being told the early results from the election. And there's even one based on the sub-prime mortgage crisis, in which Hitler is a property investor ('All over the map, *Mein Führer*, mortgages are resetting and homes are going into foreclosure'). And so on, *ad infinitum*.

Some of these parodies were tiresome. But many were side-splittingly funny – a testimony to the power of remixing as a way of enlivening cultural life. Nevertheless, not everyone was delighted by this new art form. Jewish organizations were (understandably) disturbed by the way the architect of the Holocaust had been turned into a comic turn. 'Hitler,' said the director of the Anti-Defamation League (a Jewish lobbying group), 'is not a cartoon character'. Eventually, the makers of the original film decided that they'd had enough and asked Google (which owns YouTube) to take down the remixes.[50]

To an intellectual property lawyer, this will seem entirely straightforward. To anyone interested in art and culture, however, the issue is more complicated. For one thing, there's the awkward fact that all artistic endeavour involves borrowing from earlier art works in one way or another. Just think of

Handel, whom we revere as a composer but who was a notorious, 'borrower'[51]. And every song ever written has been informed by music that the composer has absorbed in his or her earlier life.

Or, as T.S. Eliot put it, one of the 'surest tests' of the superiority or inferiority of a poet

> is the way in which a poet borrows. Immature poets imitate; mature poets steal; bad poets deface what they take, and good poets make it into something better, or at least something different. The good poet welds his theft into a whole of feeling which is unique, utterly different than that from which it is torn; the bad poet throws it into something which has no cohesion[52]

The YouTube remix culture is thus a contemporary take on an old and venerable tradition – with the twist that you don't have to be T.S. Eliot to remix something and have it seen by millions of people.

It would be silly to argue that the *Downfall* spoofs are high art. Some commentators, like Jaron Lanier, a prominent computer scientist, composer, visual artist and author, see this kind of stuff as a second-rate, parasitic cultural form. When something new is created on the Web, he told a British reporter, 'it's often a banal "**mashup**" of old culture'. He likens the people who create these mashups to 'salvagers picking over a garbage dump'. It is only in the old-world economy – the world of books, films and newspapers – that original content is being created. User-generated content is, in Lanier's view, undermining the old-world economy 'in favour of one based on free content and selling social graphs to advertisers.'[53]

But mashups are just one kind of user-generated content. And whatever one thinks of it, this explosion in peer production is one of the most distinctive features of the new media ecosystem, and we can only speculate about what its long-term impact will be. One conjecture is that it may signal a reversal in the decline of what the German cultural critic Jürgen Habermas calls 'the public sphere'[54] – an arena which facilitates the public use of reason in rational-critical debate and which had been steadily narrowing as the power and reach of mass media increased in the industrialized information economy. In recent years the political implications of this re-energized public sphere have begun to emerge, notably in elections in the US and Europe.[55]

Signs of this include the way in which political blogging played a more prominent role in the 2008 US presidential election and the 2010 UK general election than it had in earlier contests. Broadcast television was still the most dominant medium in both elections, but its influence was vitiated in interesting ways. Take, for example, the way in which Barack Obama sought to defuse the issue of race and the threat to his candidacy posed by the extremist language of his former pastor, the Rev. Jeremiah Wright (who had declared, among things, that the 9/11 attacks were 'like chickens coming home to roost').[56]

Obama's chosen method to deal with the problem was to deliver a powerful, carefully crafted speech with the title 'A More Perfect Union' in the city of Philadelphia, which most Americans regard as the cradle of their democracy. By the standards of TV-driven sound-bite campaigning, the speech was unconscionably long and discursive.[57] In the old media ecosystem – for example in the previous (2004) presidential election – it would have been reported using the standard news-bulletin template: a scene-setting opening sequence followed by a few edited highlights from the speech, plus a closing piece to camera by the reporter. For the vast majority of Americans, therefore, the burden of the speech would be hidden from them; they would have been entirely dependent on how the TV networks chose to frame and present it.

But by March 2008 something important had changed in the media ecosystem: YouTube had appeared on the scene.[58] So the Obama speech – *in its entirety*, exactly as delivered – was instantly available to anyone who wanted to see it.[59] When I last looked, it had been viewed by over eight million people. By the standards of Superbowl audiences this may not seem a big deal. But in a world hitherto dominated by sound-bites and contrived photo-opportunities, the fact that such a lengthy, complex speech should have received this much public attention begins to look pretty significant. The public sphere is expanding again.

In fact, it may be evolving into what the network scholar William Dutton calls the 'Fifth Estate' – i.e. 'a new form of social accountability' that is

> enabled by the growing use of the Internet and related information and communication technologies (ICTs), such as the personal computer and World Wide Web. Essentially, the Internet is enabling people to network with other individuals and with a vast range of information, services and technical resources. This is being achieved in ways that can support greater accountability not only in government and politics, but also in other sectors.[60]

Dutton believes that this could be as important – if not more so – to the twenty-first century as the Fourth Estate has been since the eighteenth.

Productivity 2.0

The new ecosystem may also be more productive than the old. The fact that new media have become synonymous with 'remix' culture is no accident: it's a predictable by-product of digital technology which dramatically alters the economics and reproducibility of cultural artifacts – whether they be songs, images, movies, articles or books.

Let's take economics first and illustrate the point with an example from the music industry. The economics of the compact disc (manufacture, packaging, shipping, warehousing and retailing represented about fifty per cent of the cost of a CD) meant that the most economical 'unit' of recorded music was the album; selling individual tracks using CD technology was either less profitable or downright uneconomic. But from the moment when music bitstreams could be transported across the network at very low cost, the underlying economics changed, and with them the basic unit of music – which reverted to the track or the individual song. So one implication is that it makes smaller units of 'content' economically viable because the costs of distribution fall to near-zero levels.

On the technological front, there are two factors at work.

The first is *convergence* – the fact that all media have converged into **bitstreams**. Virtually all of the 'content' that is nowadays created exists as streams of ones and zeroes. If you were to examine the contents of my computer's hard drive – or of any of Google's servers for that matter – all you would see are ones and zeroes. These digits may represent a fragment of an image, a book, a movie, a TV programme or a song, but all of those different media are now expressed in a simple, machine-readable format. So one of the barriers to re-use in the analogue world – the physical difficulty of taking something in one format and mixing it with something else – has effectively vanished. With the right software we can nowadays mix and re-use just about anything.

Secondly, digital technology transforms copying from an activity that was awkward, inefficient and degenerative (in the analogue world) into a process that is effortless, faultless (in the sense of producing perfect reproductions) and cheap. Once you have a digital original, the marginal cost of making a perfect copy is tiny.

These factors combine to create a cornucopia of new possibilities, but there is a common trend: it is that the unit of cultural transmission is inexorably becoming smaller. Instead of albums, we have tracks; instead of whole movies or TV series we have sequences (as with *Downfall*); newspaper reports are disassembled into headlines and paragraphs or RSS feeds; blogs are fragmented into individual posts; and so on. And not even books are immune, it seems. Kevin Kelly, a well-known digital 'visionary' foresees a shift from the conventional vision of the library's future which assumes that books (or even e-books) will remain isolated items, detached from one another, as they are in shelves in a public library where, as he puts it, 'each book is pretty much unaware of the one next to it':

> When an author completes a work, it is fixed and finished. Its only movement comes when a reader picks it up to animate it with his or her imagination. In this vision, the main advantage

> of the coming digital library is portability – the nifty translation of a book's full text into bits, which permits it to be read on a screen anywhere. But this vision misses the chief revolution birthed by scanning books: in the universal library, no book will be an island.[61]

As Kelly sees it, turning inked letters into electronic dots that can be read on a screen is simply the first essential step in creating this new library. The real magic, he says, 'will come in the second act, as each word in each book is cross-linked, clustered, cited, extracted, indexed, analyzed, annotated, remixed, reassembled and woven deeper into the culture than ever before. In the new world of books, every bit informs another; every page reads all the other pages.'[62]

The increasing *granularity* of online culture has stirred up a hornet's nest of concerns. For some dismayed observers it's evidence of dumbing down or of collective Attention Deficit Syndrome. Others see it as destructive of the organic unity of artistic works. The novelist, John Updike, for example, raged against Kelly's vision. 'Unlike the commingled, unedited, frequently inaccurate mass of "information" on the Web,' he told a publishing conference in 2006, 'books traditionally have edges. But the book revolution, which from the Renaissance on taught men and women to cherish and cultivate their individuality, threatens to end in a sparkling pod of snippets'.[63]

Jaron Lanier is even more scathing. Writing about the Google Books project – which aims to digitize all the world's books and is sometimes caricatured as the literary equivalent of the wartime Manhattan Project (which created the first atomic bomb) – he writes:

> What happens next is what's important. If the books in the

> cloud are accessed via user interfaces that encourage mashups of fragments that obscure the context and authorship of each fragment, there will be only one book. This is what happens today with a lot of content; often you don't know where a quoted fragment from a news story came from, who wrote a comment, or who shot a video. A continuation of the present trend will make us like various medieval religious empires, or like North Korea, a society with a single book.[64]

Like it or not, though, we seem to be heading for Updike's 'sparkling pod of snippets'. At any rate, that's the current direction of travel of both the technology and the economics of digitized information. The cultural implications of this shift are unknowable at present but may ultimately be profound – which is why the trend has already spurred Utopian dreams and dystopian nightmares. But can we say anything more useful about cultural granularity?

Actually, yes. Firstly, it's not a completely new phenomenon. In early modern Europe, the keeping of *commonplace books* (i.e. notebooks in which people recorded snippets from experience or passages from texts they were reading) was a popular obsession with readers. 'Unlike modern readers,' writes Robert Darnton, Harvard's Library Director,

> who follow the flow of a narrative from beginning to end, early modern Englishmen read in fits and starts and jumped from book to book. They broke texts into fragments and assembled them into new patterns by transcribing them in different sections of their notebooks. Then they reread the copies and rearranged the patterns while adding more excerpts. Reading and writing were therefore inseparable activities. They belonged

> to a continuous effort to make sense of things, for the world was full of signs: you could read your way through it; and by keeping an account of your readings, you made a book of your own, one stamped with your personality.[65]

In other words, early readers were doing their own version of remixing – taking snippets of other people's thoughts and making personalized mashups from them. These prolific borrowers were not the seventeenth-century equivalent of YouTube's acne-ridden teenage anarchists, either, but some of the luminaries of the Western intellectual tradition. The philosopher John Locke for example, first began keeping a commonplace book in 1652, his first year at Oxford, and over the next decade developed an elaborate indexing system for keeping track of its growing body of content. And Thomas Jefferson was an avid 'commonplacer' whose remix of the New Testament was possibly the most notorious publication he ever essayed.[66] Other practitioners of the art were Erasmus, Ben Jonson, John Milton and Francis Bacon.[67]

The fact that the Enlightenment was driven by thinkers who relentlessly and creatively borrowed ideas from one another and remixed them should serve as a useful antidote to contemporary assertions that remixing is sharply correlated with dumbing down. In fact, one could argue exactly the opposite case – that intellectual life and cultural development thrive in environments which make it easy to abstract, excerpt, borrow and remix. As Matt Ridley puts it, the rate of cultural and economic progress depends on the intensity with which ideas interact. In that sense,

> globalization and the Internet are bound to ensure furious economic progress in the coming century – despite the usual

> setbacks from recessions, wars, spendthrift governments and natural disasters. The process of cumulative innovation that has doubled life span, cut child mortality by three-quarters and multiplied per capita income ninefold – world-wide – in little more than a century is driven by ideas having sex. And things like the search engine, the mobile phone and container shipping[68] just made ideas a whole lot more promiscuous still.[69]

On this view, the argument that such a ferment should be outlawed because it sometimes involves infringement of copyright, could be turned on its head. The problem is not infringement *per se*, but intellectual property regimes that are not properly aligned with society's current needs. That is not to say that intellectual property is unimportant, or that IP rights should be violated – only that society's needs for creativity and innovation in a digital ecosystem need to be balanced against – and may sometimes trump – the interests of rights holders, and that the balance struck for an analogue age threatens our prosperity by making it difficult to harness the productive potential of digital technology.

'When text is free to combine in new, surprising ways,' says Steven Johnson, the cultural critic and online commentator, 'new forms of value are created'.

> Value for consumers searching for information, value for advertisers trying to share their messages with consumers searching for related topics, value for content creators who want an audience. And of course, value to the entity that serves as the middleman between all those different groups. This is in part what Jeff Jarvis [a prominent new media commentator, author and academic who blogs at www.buzzmachine.com and teaches

at New York University] has called the 'link economy,' but as Jarvis has himself observed, it is not just a matter of links. What is crucial to this system is that text can be easily moved and re-contextualized and analyzed, sometimes by humans and sometimes by machines.[70]

Like me, Johnson believes in thinking about our media environment in ecological terms, and he makes ingenious use of the ecologists' concept of 'productivity', i.e. a measure of how effectively an ecosystem converts input energy and nutrients into biological growth. A productive ecosystem, like a rainforest, sustains more life per unit of energy than an unproductive ecosystem, like a desert. 'We need,' Johnson thinks, 'a comparable yardstick for information systems, a measure of a system's ability to extract value from a given unit of information. Call it, in this example: textual productivity. By creating fluid networks of words, by creating those digital-age commonplaces, we increase the textual productivity of the system.'

When text is free to flow and combine, Johnson argues, 'new forms of value are created, and the overall productivity of the system increases.[71] But of course, when text is free, value is sometimes subtracted for the publishers who used to charge for that text.' And this, in a way, goes to the heart of the controversies that the rise of the Internet has inflamed. For while the *overall* productivity and creativity of the system may be increased, many of the species that have hitherto enjoyed a comfortable living will find that the environment has become inhospitable, and that they face a simple choice: adaptation or extinction.

The ecological metaphor is thus not a comforting one for established media species. It becomes even less comforting when they realize that, in the natural world, there appears to be a strong correlation between biodiversity and overall ecosystem productivity. As peer production becomes more widespread and established, for example, the number of 'publishers' expands at an exponential rate. These are not publishers in the industrialized sense of the term, of course; they are small, agile and adaptable and are not always motivated by economic considerations. But they are publishers nonetheless, and they have access to a low-cost, global distribution network. And while the things that they create, consume, publish and remix are smaller in scale than the books, movies, albums, symphonies and so on that were the stock in trade of the old ecosystem, they are nevertheless seen as desirable and are exchanged in great volumes and at a furious pace. To the lumbering dinosaurs that dominated the old media ecosystem, it must look as though their world has been taken over by termites and beetles. Small wonder that they feel depressed. Beetles make up about forty per cent of the known extant species on our planet, and they are probably the only living form that could survive a nuclear holocaust.

5. COMPLEXITY IS THE NEW REALITY

In the Beginning was the Word. So says the first line of St John's Gospel, in a useful reminder that, once, all communication was oral. Then came writing, about which Plato was famously skeptical,[1] and which was eventually systematized into the production of manuscripts by specialized scribes. Then Gutenberg exploded onto the scene, and the printing press became the main engine of communication, spawning first books and pamphlets and, later, newspapers and magazines.[2] The arrival of universal postal services made possible the rapid and reliable circulation not just of printed works, but also of private communications. After that, in 1843, came the first of the electronic media – the telegraph – memorably christened by Tom Standage as 'the Victorian Internet',[3] followed by the telephone (1876) and the phonograph (1877). The closing years of the nineteenth century saw the spread of mass literacy followed by the emergence of the first 'popular' (i.e. mass-circulation) newspapers[4]. Then came movies and the broadcast media – first radio in the 1920s and television in the late 1940s.

Each of these new technologies changed the media ecosystem of which they became components. Looking back on the history, one clear trend stands out: each new technology increased the *complexity* of the ecosystem. The arrival of the Internet, and particularly of the Web, has imparted a dramatic twist to the complexity spiral, in the sense that the media ecosystem that is emerging as a result of the network is orders of magnitude more complex than anything we have seen before. And the implications of this are just beginning to dawn on us.

Unpacking complexity

'Complexity' is one of those deceptive terms that we use all the time while assuming that everyone knows what it means. It's regularly employed, for example, as a synonym for 'difficult', 'intriguing', 'incomprehensible', 'complicated' and 'mysterious'. So, wine lovers talk approvingly of the complexity of a vintage claret; mathematicians revel in the complexity of an abstruse proof; software engineers bemoan the complexity of the computer code that other programmers wrote and they are now expected to maintain. Urban planners proudly outline plans for a new 'sports complex', while psychologists talk of clients suffering from an 'inferiority complex' and mathematics teachers struggle to explain 'complex numbers' (which contain a real and an imaginary part) to baffled pupils. And so it goes on.

Underpinning this linguistic casualness are two basic conceptions of complexity. One is *subjective*: we say that something is complex if we are unable to understand it; in that sense a complex system is one for which we feel that our mental model of it is inadequate. The other conception is what mathematicians call *objective* complexity – i.e. where a system has properties (see later) that entitle it to be regarded as complex. But in one respect at least, it doesn't matter which interpretation of complexity we endorse. The fact is that complexity exists, and is often the bane of our modern lives.

But what is it, exactly? Given that the last few decades have seen a massive expansion in the academic discipline variously known as 'complexity science' or 'the sciences of complexity', you'd have thought that a definition would be easy to come by. Amazingly, it isn't; indeed there appears to be no widely agreed comprehensive definition.[5] So I've gone back to one of the seminal texts in the subject – an article on 'The Architecture of Complexity' published nearly half a century ago by the Nobel Prize-winning economist, Herbert Simon. 'By a complex system,' Simon wrote,

> I mean one made up of a large number of parts that interact in a non-simple way. In such systems, the whole is more than the sum of the parts, not in an ultimate, metaphysical sense, but in the important pragmatic sense that, given the properties of the parts and the laws of their interaction, it is not a trivial matter to infer the properties of the whole.[6]

This property – of the whole being in some sense more (or less) than the sum of the parts – is called *emergence*. (A simple illustration: the pungent smell of ammonia, which is a compound of two odourless gases – hydrogen and nitrogen. You could say that the distinctive smell is an *emergent property* of the molecule.)

In addition to emergence, complex systems are found to possess some or all of a range of properties which make them hard to understand or predict.[7] For example:

- They are *non-linear* – which means that small changes to one part of the system may produce disproportionately large changes in other parts, or in the behaviour of the system as a whole. Worse still, complex systems can exhibit what mathematicians call *chaotic behaviour* – by which they mean that small differences in their initial conditions can yield widely diverging outcomes – the so-called 'butterfly effect'.[8] This terminology is confusing to non-mathematicians because it differs from the everyday interpretation of 'chaos' as a state of extreme confusion and disorder, but we're stuck with it.[9] It's important because what it means in practice is that complex systems may be intrinsically unpredictable.
- *Feedback* plays an important role in determining their behaviour. This can be of two kinds – positive, in which a change in some state of the system stimulates further change which in turn leads to even more change in a self-reinforcing, runaway pattern leading to exponential growth; and negative feedback, in which changes stimulate smaller changes leading to a new (lower) equilibrium – rather as a national economy in recession may stabilize at a lower level of activity.
- *Self-organization*. Some systems exhibit spontaneous order – i.e. order that arises from the aggregation of very large numbers of uncoordinated interactions between their components. The behaviour of ant colonies is a classic example. If the apparent contradiction between 'chaos' and the appearance of spontaneous order bothers you, remember that mathematicians use the term 'chaos' in a specialized sense; and also that not all complex systems exhibit chaotic behaviour.

How does one measure complexity? Just as with the definition of the concept, there seems to be no general agreement on metrics. Ecologists measure complexity using one set of measures; physicists use others; social scientists have their own metrics, as have meteorologists, thermodynamicists, architects, urban planners, computer scientists, biologists, engineers, economists, game theorists, and all the other people who study complex systems in their own specialist fields. Critics of complexity science see this profusion of measures as evidence of confusion. A more sympathetic observer, Seth Lloyd, interprets it as a sign that, as an academic discipline, the study of complexity is still relatively young and immature. As he sees it,

> the problem of measuring complexity is the problem of describing electromagnetism before Maxwell's equations. In the case of electromagnetism, quantities such as electric and magnetic forces that arose in different experimental contexts were originally regarded as fundamentally different. Eventually it became clear that electricity and magnetism were in fact closely related aspects of the same fundamental quantity, the electromagnetic field.[10]

So if contemporary researchers in complex systems tend to come up with different measures of complexity, it may simply be a reflection of the fact that the field still lacks an overarching theory – or its own James Clerk Maxwell.

Complexity and our media ecosystem

In the absence of a contemporary Maxwell, how might we assess the complexity of our emerging, Internet-dominated media ecosystem? My suggestion is that we take three simple, common-sense measures, the first two of which are inspired by Herbert Simon's original, 1962, definition. Here is the list:

- The number of components – that is, agents or participants involved as consumers and producers of content
- The density of interconnections and interactions in the system
- The speed of change and development within the ecosystem.

On each of these measures, the complexity of our current ecosystem is significantly greater compared with the system as it was before the Internet went mainstream.

Number of components

Prior to the Internet, the media ecosystem was a relatively simple system, with a limited number of creators and publishers of content communicating it to mass audiences of readers, listeners and viewers who were for the most part seen as isolated and passive consumers. The system was largely one-way, dominated by push media like broadcast TV. This system of 'mass media' was a product of economic and technological forces that had become dominant in the late nineteenth and early twentieth centuries. As Yochai Benkler put it,

> The core distinguishing feature of communications, information, and cultural production since the mid-nineteenth century was that effective communication spanning the ever-larger societies and geographies that came to make up the relevant political and economic units of the day required ever-larger investments of physical capital. Large-circulation mechanical

> presses, the telegraph system, powerful radio and later television transmitters, cable and satellite, and the mainframe computer became necessary to make information and communicate it on scales that went beyond the very local. Wanting to communicate with others was not a sufficient condition to being able to do so. As a result, information and cultural production took on, over the course of this period, a more industrial model than the economics of information itself would have required.[11]

Two technological developments changed this: the affordable personal computer together with its accompanying software, and the Internet, which enabled these devices (and their owners) to communicate and publish. The result was that the physical requirements for effective information production and communication were now 'owned by numbers of individuals several orders of magnitude larger than the number of owners of the basic means of information production and exchange a mere two decades ago'.[12] The scale of this transformation has been of truly epic proportions. As I write, the standard estimate of Internet users is around two billion people.[13] Or, to put it another way, almost a third of the world's population now has an Internet connection of some sort. On any given day users upload millions of digital images to hosting and social-networking sites like Flickr.com and Facebook, write (and read) over 100 million blogs and upload an astonishing amount of video footage to YouTube.

We've moved into an era characterized by the cultural critic Clay Shirky as the 'mass amateurization of publishing'[14] – or what Yochai Benkler calls 'social production' to capture the fact that whereas, in the old ecosystem, most publishing took place with a market framework, now a great deal of it also happens outside the marketplace. The point is not that one has replaced the other, but that the number of independent agents in the publishing space has expanded enormously and that fact, by itself, makes the ecosystem more complex.

Density of interconnection and interaction

In the pre-Internet ecosystem there were fewer channels for communication and fewer means for individuals and organizations to interact with one another. Before 1960, for example, most of the world's population had probably never made a phone call, and even in the industrialized world many households – and most young people – did not possess a telephone. Interactions were limited mainly to face-to-face encounters, telephone calls and letters.

Compare that with the situation today in which people exchange billions of email, text and instant messages, post hundreds of millions of status updates to social-networking sites, and write millions of blog posts – all on a daily basis. We've created a world in which a blizzard of RSS feeds provides an automated update stream to subscribers around the clock, in which a celebrity Twitter user like the British actor Stephen Fry can instantly flash a message to his 2.7 million 'followers'[15] and in which even traditional few-to-many broadcasters now routinely invite audience participation via text, email, instant messaging and Twitter – as well as by means of old-fashioned voice calls. This is a world in which teenagers in many countries use social networks and text messaging to organize parties and meetings, in which political activists use Facebook, SMS and Twitter to organize demonstrations, protests and marches, and in which pranksters use the same technologies to mobilize 'flash mobs'.[16]

The intensity of interactions in the new ecosystem is vividly exemplified by what goes on in one social-networking site, Facebook. According to one report,[17] in an average twenty-minute period in 2010 users of the service engaged in the following activities:

- Shared 1 million links
- Tagged 1.32 million photographs
- Wrote 1.59 million wall posts
- Uploaded 2.72 million photographs
- Wrote 10.21 million comments
- Sent 4.63 million messages

Speed of change

The third dimension of complexity is the speed at which things change – the pace of development, the rate at which new system components emerge and grow. In this respect too the new ecosystem differs radically from the old. Just think: none of the nine companies and organizations that nowadays impact most heavily on the daily lives of millions of Internet users (Amazon, Google, eBay, Yahoo, YouTube, Facebook, Skype, Wikipedia, Craigslist) existed twenty years ago. As I write this, the oldest (Amazon and Yahoo) are just sixteen years old, while the newest (YouTube and Facebook) have only recently passed their respective fifth and sixth birthdays. In its brief lifetime, Facebook has gone from a standing start to over 600 million users.[18] And Facebook is not alone in this: Skype – the Internet telephony company – also has over 500 million registered users.

The history of capitalism is littered with episodes in which companies appeared and grew, so a sceptic might argue that the only thing that's remarkable about the Internet-powered growth spurt of the last two decades is the speed with which it has happened. But, as far as complexity is concerned, the speed is precisely the point. For while some of these new enterprises have created entirely new fields of activity – things which hadn't been possible before the arrival of the network – some of them have had a major impact on existing industrial structures and incumbents. Think, for example, of how Wikipedia has affected the market for printed works of reference; of how Amazon and eBay have affected offline retailing; of how Google News has affected the newspaper business; of how Craigslist undermined the offline classified advertising business; and of how Skype, by enabling its users to make free phone calls over the Internet, is having an impact on traditional telecoms companies.

Taking complexity seriously

It's clear that, even on the basis of the three obvious measures that I've used, our emerging media environment is not only a complex system, but one whose complexity is significantly greater than that of its predecessors. For many readers, this won't come as a shocking revelation. 'Of course it's complex,' they will say, 'tell us something new'. For them, it's akin to observing that the English weather is changeable – a fact of life, but one that we can't do much about. It may influence our day-to-day behaviour, but won't have much effect on our long-term strategies.

You won't be surprised to learn that I don't see things that way. For me, the complexity of our media environment is not a banal truism, but a reality that needs to be taken seriously by anyone who intends to operate in this ecosystem. A better analogy might be to say that its formidable complexity is akin to a speeded-up version of global warming – that is to say, a reality that will one day affect everyone and which demands radically new thinking now.

What kinds of new thinking? Well, first of all, we need a change of attitude. Just as anyone who seeks to understand the Net has to realize that disruption is a feature of the system, not a bug, so we need to accept that complexity is something we have to live with. It's not a temporary aberration, but the new reality. And it's likely to increase.

Secondly, we need to accept that although complex systems are difficult to understand, it *is* possible to study them, to model them and to think intelligently about them. Just as financial institutions had to recruit mathematicians and engineers to enable them to cope with the complexity of a deregulated market for financial products and services, so media organizations need to tool up intellectually – to recruit new kinds of expertise beyond that of the lawyers, accountants and market-researchers to whom they have traditionally outsourced their thinking.

Most importantly, organizations will need to respond to the most important single implication of operating in a complex system, namely that its behaviour is intrinsically unpredictable. The strange thing about most corporations (and other large non-commercial organizations, including universities) is that although their leaders all pay lip-service to the proposition that the future is unknowable, in practice they have created institutions that are structured on the assumption of predictability. Taking complexity seriously therefore means thinking hard about what kinds of organizations – and managerial mindsets – are needed.

Some lessons for management from complexity theory were elegantly summarized by the LSE management scientist, Jonathan Rosenhead. 'The general lessons,' he writes, 'concern how learning can be fostered in organizations, how they should view instability, and the (negative) consequences of a common internal culture'.

The need for an emphasis on learning stems from the central finding of this theory – that the future is in principle unknowable for systems of any complexity. If we accept that we can have no idea of the future environment, then long-term planning becomes an irrelevance, if not a hindrance. The absence of any reliable long-term pattern makes what is called 'double-loop' learning important. That is, it is not enough for managers to adjust their behaviour in response to feedback on the success of their actions relative to pre-established targets; they also need to reflect on the appropriateness, in the light of unfolding events, of the assumptions (the mental model) used to set up those actions and targets.[19]

Rosenhead believes that this sort of learning cannot easily take place within an organization which – as most contemporary organizations do – puts a premium on maintaining a common culture. The dynamics of 'group think' and the possible effects of divergent views on an employee's promotion prospects strongly encourage conformity. This is not an atmosphere in which searching re-examination of cherished assumptions can thrive – rather the reverse. Yet agility of thought based on the fostering of diversity is a prerequisite for the organization's longer-term success.

The implications of complexity are that organizations which

seek stable equilibrium relationships with an environment which is inherently unpredictable are heading for failure. Successful strategies, especially in the longer term, will not result from fixing an organizational strategy and mobilizing around it. What will be important is openness to change, accident, coincidence, serendipity.

Look around our established media organizations and ask if any of them possess the characteristics identified by Rosenhead as necessary for survival in a complex ecosystem. And then consult your broker.

6. THE NETWORK IS NOW THE COMPUTER

Sometime in the early 1980s, a computer scientist named John Gage said something that turned out to be very profound: 'the network,' he declared, 'is the computer'.[1] To most people at the time that must have seemed an incomprehensible statement, for two reasons. Firstly, simple local-area networks (LANs) which connected PCs to printers and other shared facilities were still relatively exotic; and secondly, the Internet as we know it today had only recently been switched on – a development of which the world beyond a few computer science labs was blissfully unaware.

Incomprehensible or not, Gage's aphorism turned out to be prophetic. We've now reached the point where it has become an accurate description of the world we inhabit. It's taken a while to get here, and the route has been tortuous at times. And the implications of Gage's insight are far-reaching. But before we consider them, we need to understand the journey we've travelled.

How we got here

Let's start at the beginning. What is a computer?

Until relatively recently, the word referred to a *person* –

someone who did calculations on behalf of someone else. That's the sense in which astronomers of the eighteenth and nineteenth centuries would have used the term: it described the legions of people they employed to work out the orbits of planets or the distance of stars or a thousand other empirical facts they wished to establish on the basis of their theories or observations.

In my lifetime, the term computer has come to mean an electronic machine which does calculations or 'computations'. It does things with (or, in mathematical terms, performs operations on) numbers, specifically ones and zeroes. Strange as it may seem, operations on these binary digits underpin everything a computer does – even when what it appears to be doing is placing these letters on a screen in response to the keystrokes I'm making on its keyboard.

So, for us, a computer is a machine. Such machines have their origins in the Second World War, specifically in the code-breaking efforts of Alan Turing and his colleagues at Bletchley Park, the top-secret establishment located midway between Oxford and Cambridge in which the British broke the German Enigma code (among others), thereby gaining a decisive advantage in the war against Hitler.

The early computers were colossal machines with thousands of thermionic valves, switches and dials. They took up entire rooms, generated vast amounts of heat and broke down frequently. But as electronic engineers made the transition from valves to silicon-based transistors,[2] the amount of heat generated declined and the reliability of computers increased. But they were still vast in size, fabulously expensive to build and required an extensive 'priesthood' of specialist staff to program and operate them. This perhaps explains why in 1943 Thomas

Watson, the head of IBM, allegedly estimated the total world market for computers at about five machines.[3]

One of the reasons the view attributed to Watson was wrong is that it underestimated the variety of uses that corporations and governments would find for general-purpose computers. The giant corporations of the post-war world processed vast amounts of information – invoices, orders, reports, inventories – and employed millions of people to do this processing. But people are expensive; they have to be recruited, inducted, trained, managed – and paid. Much of the information-processing they did was routine, predictable stuff. If these tasks could be automated – performed by a machine – then the attractions of the new technology were obvious. And even if the machine appeared at first sight to be horrendously expensive, the benefits – in terms of reduced cost, improved reliability, consistency and speed – greatly exceeded the costs of acquiring and operating a data-processing machine.

Thus was born the 'mainframe' computer. Companies came to assume that running a complex computing operation was an intrinsic part of doing business. By the mid-1950s, the corporate mainframe was an established technology for doing routine 'data processing'. But then companies began to realize that computers might also give them strategic competitive advantages over their competitors. In 1959, for example, American Airlines started work on a computerized system for making – and confirming – ticket reservations in 'real time'. By 1969 its SABRE system was up and running, and its competitors found themselves with no option but to follow suit.[4] Much the same happened in banking, with Bank of America leading the way. And where the airlines and the banks went, eventually the rest of the corporate world followed, with the result that mainframe computers became the dominant technology of the business world.

These developments were mirrored in the areas of defence and higher education. The nuclear standoff between the US and the Soviet Union which characterized the Cold War put a very high premium on being able to detect missile launches in order to activate retaliatory action in 'real time', and so mainframe computers were seen as an essential component of a nuclear war-fighting capacity.[5] And in universities they were seen as essential tools for research, as well as for training and educating successive generations of engineers. So by the 1960s, these huge, expensive machines were everywhere in the organizational world.

The formidable costs involved in mainframe computing determined how they were used. It was important, for example, that such expensive assets should be worked hard. The typical corporate mainframe, according to Nicholas Carr, operated at more than ninety percent of its full capacity.[6] This was achieved by what was called 'batch processing': programmers encoded their work on decks of punched cards or rolls of paper tape and submitted them to operators who would schedule their processing to fit in with the rest of the machine's work-flow. If your job had high priority, then it was run quickly; if it didn't, then it just had to wait in line. Low-priority jobs would be run in the middle of the night.

Universities and research institutes initially operated their mainframes in batch-processing mode, although this was frustrating for academic users. Their frustrations were eased by innovations in operating systems that enabled these mainframes to become 'time-shared' machines. Dozens – eventually hundreds – of users could log on to the mainframe, write, compile and run programs – and see the results immediately or shortly afterwards. Each user had the impression that s/he had the machine's full attention, but in fact this was an illusion fostered by the operating system, which was essentially cycling rapidly round each user, allocating her or him milliseconds of processing time.

Time-shared computing was a clever technology, but underpinning it were some harsh realities. The mainframe might have appeared 'personal' to its users, but that was an illusion. In fact the computer was a resource controlled and administered by others who determined who could become an authorized user, how much processing time and disk space one could have, and so on.

For most of my time as an engineering student, this is what 'computing' was like. In order to use a computer, one had to have an account on a time-shared machine, with which one communicated via a chattering teletype or, later, a terminal consisting of a screen and a keyboard. But during the late 1960s and early 1970s things began to change. Moore's Law – the observation made by Gordon Moore, one of Intel's co-founders, that the number of transistors that could be fitted onto a silicon chip (and therefore the chip's processing power) doubled every eighteen months – meant that computers became smaller. Time-shared machines shrank from taking up a large room to being 'minicomputers' the size of a large domestic refrigerator. And they became cheaper, so that whereas once a university could afford only a single mainframe, individual departments could afford to purchase and operate a minicomputer. To us students, these seemed a great advance on the behemoths of old. But they still had some of the disadvantages of the *ancien régime*. You still had to have permission to have a user account, for example; you were given a standard allocation of hard disk space, and if you wanted more you had to go cap in hand to the machine's administrator; and so on.

All of this changed in 1975, when an unknown company in the US which had hitherto specialized in making electronics kits for rocket enthusiasts launched the Altair 8800, the first truly 'personal' computer.

Although its designers described it as a 'minicomputer', in fact the Altair was nothing like the minicomputers of the day and was rapidly rechristened a 'microcomputer'. The machine was marketed in two forms – as a kit of parts (costing $439) that

Figure 6.1: The cover of the January 1975 issue of Popular Electronics. (source: http://en.wikipedia.org/wiki/Altair_8800)

customers could assemble for themselves; and as a pre-assembled product costing $621. Its manufacturers expected to sell 800 machines, but by February 1975 already had orders for 1,000. The launch of the machine was what spurred a young Harvard undergraduate named Bill Gates to drop out of college: he intuited that the Altair was the beginning of something big and wanted to be part of it.[7]

In that, at least, Gates's judgement was sound. The Altair was the prototype of a new kind of computer – one that was truly *personal*, in that it belonged to its owner. Up to this point, anyone who wanted to use a computer had to be granted an account on a mainframe or a minicomputer, and had to accept whatever restrictions were imposed as a condition of admission. Now, you could be master of your own machine. Granted, it was a pretty primitive device, limited in the strength of its processing power, memory size, storage capacity and peripherals like printers. But it was yours to do what you liked with.

And it was expandable. The Altair was, in its architecture, a model of things to come. This model was a modular design in which the various essential components – processor, memory, disk drives, interfaces for peripherals – were separate units mounted on printed circuit boards (called 'cards') linked by a common electronic connector called a *bus* which played a central role in the machine.[8] When the processor needed to fetch data from memory it sent the request via the bus, and the desired data travelled back along it. Similarly, if data had to be read from, or written to, a floppy or hard disk, the requests and the data travelled back and forth along the bus. In that sense, a personal computer is like a busy city with only a single paved road – the bus.

As often happens with pioneers, the company that made the

Altair did not become a dominant force in the industry that it spawned. It was companies like Apple, IBM, Microsoft, Compaq, Dell and Hewlett-Packard that reaped the reward of the Altair's ingenuity. But the Altair's inheritance shaped the mindset with which we approached computers for the next twenty-five years – and which still blinds us to the significance of what has happened since the Internet became central to our lives.

This mindset can be brutally summarized in a single, elemental belief – *that the PC is the computer*.

In the first two decades of personal computing, this was not only a perfectly understandable belief: it was also true. If you wanted to do 'computing' – i.e. avail yourself of computing services like word-processing, desktop publishing, email, spreadsheet modelling, database work, image manipulation, document printing – then you had no option except to power up your PC and launch the appropriate program(s). Everything that happened as a consequence happened within your machine, whose various components engaged in frenzied **bit**-trafficking over the machine's bus. The PC really was the computer.

But then, as the world tiptoed into the new millennium, all that began to change. The change was experienced differently by individuals and corporations, so the story forks for a bit. But in the end both routes converged on the same place – the world envisaged by John Gage all those years ago.

The user's journey

For individuals, the story starts with email – the Internet's first 'killer app'. Up to the mid-1990s, anyone outside of the world of mainframe computers who wanted to send and receive email did so via a special program (called a 'client' in computerspeak) that ran on her or his PC.

This was fine for people operating from home, but it had significant downsides for employees who were compelled to use their organizations' computers for email when they were at work. Among other things, their employers were able to monitor all email messages sent 'on company time' or via its networks; and many companies specifically prohibited workers from using company email for personal purposes.

Silicon Valley legend has it that this degree of employer surveillance was particularly irksome to geeks, as it made it difficult for them to look for work in other companies during office hours. As the first Internet boom got under way in 1994, there was a hiring frenzy in the Valley – which meant that anyone with software engineering skills effectively found themselves in a seller's market. In such circumstances, companies took exception to employees using their email systems to send out job applications and resumes (CVs) to competitors. And the legend says that this is what motivated two ex-Apple engineers, Sabeer Bhatia and Jack Smith, to invent a completely different way of doing email. Initially, they called it HoTMaiL – with the capitalized letters (HTML) spelling out the acronym for HyperText Markup Language, the formatting language used for creating web pages, but it rapidly became known simply as Hotmail.

To anyone accustomed to a PC-based email client, Hotmail was a revelation. It was a pure Web application. To access it, all you needed was a PC – any PC – which had an Internet connection and a web browser like Mosaic, Netscape or Internet Explorer (the dominant browsers of the time). You pointed the browser at www.hotmail.com and this took you to the mother ship, somewhere out on the Internet. There you could register for a free account (joesoap@hotmail.com), or sign in if you already had one. Upon logging in, your mail was displayed in the browser in a tidy list. Click on an item and it opened up for reading. Another click and a form opened up for your reply. Once you had finished typing, a click on 'Send' dispatched it on its way. And then you logged out, and nobody was any the wiser about what you had been up to.

Hotmail was launched on Independence Day 1996, and spread like wildfire. In part this was due to a clever marketing trick dreamed up by Bhatia and Fisher: the system automatically added a footnote to every email revealing that the message had been sent by Hotmail and encouraging the recipient to get a free Hotmail account at www.hotmail.com. This wheeze was later dignified by the term 'viral marketing'. But mainly Hotmail spread because it was such a lovely, simple idea. It was such a good idea, in fact, that Microsoft bought it for $400 million in December 1997, and it now lives on as Windows Live Hotmail.

The second pivotal point in the journey of individual users towards the world envisaged by John Gage came in December 1995, when a new kind of search engine appeared. It was called AltaVista and it was launched by researchers at the Western Research Laboratory of the Digital Equipment Corporation (DEC). The Web had gone mainstream in 1993 with the launch of the Mosaic browser, it began to enjoy explosive growth from 1994 onwards as it became obvious to an increasing number of people and institutions that it was a heaven-sent publishing system. But the flip side of that explosive growth was that the problem of finding anything on the Web was also going to expand at the same rate. In 1993 the search challenge could be likened to that of finding a needle in a smallish haystack; by

1995, it was like trying to find a needle in several hundred thousand haystacks.

Computer scientists had long been aware of this, of course, and even in the pre-Web era there were search facilities of various kinds. Most of them were modelled on a method of organizing information retrieval that had its roots in the print era – the catalogue or directory. But as the Web took hold of the world's imagination, it became clear that any approach based on human-compiled directories was destined to be overwhelmed. This would be a job that could only be done by computers, and so a canonical model of what became known as a 'search engine' began to take shape. It had two components: a *crawler*, i.e. a software robot or 'bot' which roamed the Web, crawling over every web page it found and logging the content; and an *indexer* which compiled a searchable database of what the crawler brought back. The early 1990s triggered a frantic outbreak of academic research and software development aimed at improving both components – making crawlers more comprehensive and efficient, and improving indexing so as to return faster and more relevant replies to search queries.

What made AltaVista special was not only the fact that it outshone its competitors on three essential criteria for a search engine – comprehensiveness, speed and relevance of results – but also that it was the first to allow natural language queries, provide advanced searching techniques and permit users to add their own URL to (or delete it from) the index within twenty-four hours. It even allowed inbound link checking and had (initially at least) a simple and uncluttered interface.[9] As a result, from the moment it appeared until Google launched in 1998, AltaVista was the search engine of choice for millions of Web users.

The significance of Hotmail and AltaVista is that they were the services that took most users over the threshold into a new world. In both cases they provided computing facilities – email and search – that users craved. But the computing involved in providing them did not happen on the users' PCs: instead it was done by servers located elsewhere – perhaps on the other side of the world. The Internet effectively replaced the PC's internal bus as the main channel for data traffic, and the PC was reduced to being a life-support system for a web browser. The significance of the shift was probably lost on most users at the time, but it meant that millions of people who thought of themselves as PC users were in effect metamorphosing into users of what Nicholas Carr later dubbed 'the World Wide Computer'.

As more and more people got broadband connections to the Net, the trail blazed by webmail and search developed into a superhighway. Today we see the results in the routine way that millions of people store and manage their digital photographs online using services like Flickr.com (which held over five billion images in late 2010) and Picasa or social-networking services like Facebook (to which its members had posted ninety billion images by 2011). We see it also in the way people use blogging services like Blogger, Typepad and Wordpress, upload videos to YouTube, watch post-broadcast TV programmes using the BBC iPlayer and listen to music via streaming services like Spotify. And even tasks like spreadsheet modelling, word-processing and presentation preparation that once required specialized software running on a PC are often now done using Web-based services like Google Docs and Zoho – again because the speed of broadband connections has made it feasible to do these things at a distance. So the Internet, as Nicholas Carr puts it, has become, literally, our computer. It's as if the various components that used to be enclosed in the metal box of the PC – the hard drive for storing data, the processor chip, the applications for manipulating data – have been dispersed and integrated via the Internet so that they can be available to everyone. In that sense, he maintains, the World Wide Web has indeed morphed into 'the World Wide Computer'. [10]

The corporate journey

The forces that propelled individual Internet users over the threshold into John Gage's imagined world were a combination of convenience and necessity: webmail was easy and free – as were most of the later web services, at least in their non-premium versions – and search could only be done as a network-mediated service.

In contrast, the logic that drove companies to rethink the way they provided computing services was mainly economic. In his account of the phenomenon, Carr sees a clear analogy between IT services and electrification. Once upon a time, every factory generated its own electricity. But eventually manufacturers realized that they didn't have to be in the power-generation business any more: they could buy electricity on a pay-per-use basis. What made this possible was a series of scientific and engineering breakthroughs – in electricity generation and transmission as well as in the design of electric motors. Electrical utilities achieved economies of scale in power production that no individual factory could match. And it was a virtuous circle. As soon as they began supplying electricity to factories, they were able to expand their generating capacity even further, achieving another leap in efficiency. The price of electricity fell so quickly that it soon became possible for nearly every business and household in the country to afford electric power.[11]

The key factor, though, was the evolution of the electricity grid – a way of transmitting electrical power from where it was generated to where it was needed.

Carr thinks that something similar is happening with computing. As we saw earlier, from the mid-1950s onwards companies had to accept that massive investment in computing and IT was an unavoidable cost of being in business. And, if anything, this overhead increased as computing technology evolved from central mainframes to huge corporate networks of desktop PCs, servers, printers and all the other paraphernalia of modern information technology. The result is that every large organization acquired a significant new overhead (i.e. cost centre). It came to be known as 'IT support'.

For the best part of four decades, that was the way things were. But as the Web began to grow from the mid-1990s onwards, two significant developments occurred. The first was the appearance of the vast 'server farms' that were needed to support large Internet Service Providers (ISPs), webmail providers like Microsoft, search engines like Google and online retailers like Amazon. The second was the massive investment in broadband capacity that accompanied the first Internet boom that mushroomed from 1995 to 2000. Suddenly the world found itself with two new assets: companies possessed of vast reserves of computing power and online storage; and a global network of high-bandwidth communication channels. A grid, if you like.

So, says Carr, what happened to power generation a century ago is now happening to computing. Company IT systems are being supplemented – or even supplanted – by computing services provided over the Internet by distant data-processing plants.

Computing, in other words, is turning into a utility. And that, in turn, will change the economics of the way companies work.[12]

In the end, a portmanteau term had to be found to encapsulate these developments. The term that emerged came from a diagrammatic convention often used by geeks when illustrating some point about web services. They would draw the Internet as a cloud.

And so it was that the world envisioned by John Gage way back in the 1980s – the world in which the network was the computer – came to be known as *cloud computing*.

Technically, cloud computing is defined as 'a model for delivering on-demand, self-service computing resources with ubiquitous network access, location-independent resource pooling, rapid elasticity, and a pay per use business model'.[13] Three factors have combined to make it one of the most significant developments of the last decade:

- On the *supply* side, large e-commerce companies like Amazon invested heavily in large server farms and related infrastructure designed to support their own core business activities. Because of the need to cope with surges in demand, this computing infrastructure had to be designed with significant amounts of surplus capacity. This meant that, in accountancy terms, Amazon found itself with assets which, for much of the time, were inevitably under-utilized. There was therefore a good business case for seeking ways of working these assets harder.
- Also on the supply side, a new technology called *virtualization* emerged. The term implies the creation of a virtual – as distinct from physical – version of something. In a computing context it means using software to create a complete, working

simulation of a computer which can be run on an actual machine. So, for example, my Apple Macintosh laptop has special software which enables me to install what is to all intents and purposes a PC running the Windows 7 operating system; it's called a 'virtual machine' but it's as if my Mac had been transformed into a Windows computer. The same software enables me to install a second virtual machine – this time a computer running the Linux operating system. And I can switch freely between the actual Mac operating system and the other two virtual machines. If my computer had enough RAM it could run more of these virtual machines. Virtualization is a technology that vastly increased the potential of the server farms run by Amazon *et al*, because it meant that each physical server in a farm could magically expand into several computers simply by hosting a number of virtual machines. This provided server farms with a hitherto-unknown level of flexibility and scalability.

- On the *demand* side, established companies were seeking ways of reducing the costs of running their own, premises-based data centres, or of finding less capital-intensive ways of meeting surges in demand. In a sense, they were just seeking a way of travelling the well-trodden path from the expense and hassle of owning, maintaining and upgrading one's own kit, to leasing or renting services from a specialist provider. And start-up companies (like Twitter) developing Web 2.0 applications were seeking ways of obtaining the computing power and storage capacity they needed without incurring the capital costs involved in setting up their own server farms.

Amazon is perhaps the best-known provider of cloud computing services. In March 2006 it launched the Amazon

S3 (Simple Storage System) which provides effectively unlimited disk storage via a web interface. For this, Amazon charges 15 cents per gigabyte per month, plus a charge for bandwidth used in sending and retrieving data, and a per-request charge. Users can specify where the server farms they wish to use are located (to reduce latency – i.e. network delay – or to comply with data-protection regulations in the jurisdiction in which they are based). The S3 service is rapidly scalable – it can cope instantly with demands for more storage – and has very high reliability and availability.

In August 2006, Amazon launched an even more significant service – its Elastic Compute Cloud (EC2). This is basically pay-as-you-go computing, allowing customers to rent virtual computers on which they can run their own applications. There's a web interface that allows customers to boot a virtual machine – or machines – running whatever specialist software their applications require. Any number of virtual machines can be launched and shut down as desired – hence the term 'elastic' – with users paying by the hour only for active machines.

Living in the cloud

What will a world dominated by cloud computing be like? The answer – as usual – is that it's too early to say, because we've only recently begun to experience it, but from what we know already we can draw some inferences about the implications of John Gage's vision. Here are some of the more obvious ones.

For users . . .

Cloud computing means that they will increasingly keep their data – blog posts, email, photographs, documents, address books, etc. – on servers located somewhere on the Internet, and access them via whatever device they happen to have at hand. The device might be a PC or a laptop; but it might just as easily be a mobile phone, an iPad or other kind of electronic 'tablet', an e-reader like the Amazon Kindle, an Internet-enabled TV set or some other as yet undreamt-of device.

An interesting question is what will happen to recorded music in a cloud-dominated world. Some years ago the singer David Bowie, while speculating about how people would 'consume' recorded music in the future, made a perceptive remark to a *New York Times* reporter. Music, Bowie said, 'is going to become like running water or electricity'.[14] To appreciate the significance of this, we need to note that he was speaking in 2002, a year after Apple had unleashed the iPod on an unsuspecting world. At the time, millions of people were transfixed by the idea that they could carry their entire music collections around with them in a tiny device. But Bowie perceived that this blissful state might just be transitory – that iPod users were, in fact, the audio equivalent of travellers who carry bottled water because public supplies are unreliable or unsafe in some countries. But once music lovers had confidence that they could always access their music online, his assumption was that they would become less concerned about carrying it round with them. The rise of on-demand music streaming services like Spotify, Amazon's Cloud Player and Apple's iCloud suggests that this change is on its way.[15]

Cloud computing also means that the devices on which people do 'computing' don't have to be PCs or laptops. In principle, they no longer have to purchase, install, maintain and upgrade specialized software to do email, word-processing, spreadsheet calculations, prepare presentations, edit photographs, audio and video: all of these tasks can now be done using cloud-based services. (I say 'in principle' because the usability of cloud-based services like the ones I mention is critically dependent on the availability, speed and reliability of the user's Internet connection.) Nor do they have to worry about having enough hard-disk storage on their machines, or about backing up their data, since data held in the cloud is automatically backed up. The implication is that users can get by with simpler, cheaper devices than a desktop or laptop computer – with so-called 'netbooks', for example; or with tablet computers or high-end mobile phones like Apple's iPhone or handsets running Google's Android mobile operating system.

So for consumers, cloud computing seems to have lots of silver linings: plenty of 'free' services available around the clock and accessible through simple, cheap devices with the only cost being the cost of a fixed or mobile Internet connection. It sounds too good to be true – and it is. All these 'free' services have to be paid for, somehow. Server farms are expensive to run; the software engineers who create and maintain the 'free' services offered by Google, Facebook, Microsoft et al. have to be paid. So a variety of business models have evolved to finance this new ecosystem.

At one extreme are services that are free to the consumer at the point of use (e.g. webmail) but which are financed, in one way or another, by advertising. At the other extreme are services that users pay for either by subscription or on a usage basis. In between are endless variations on the 'freemium' theme – i.e. a spectrum running from basic services that are free, to increasingly elaborate 'premium' services.[16]

But even if users of free cloud services are not charged a cent, no matter how much they use them, they are unwittingly paying by surrendering some of their privacy to the companies that run them. Most consumer cloud services require you to entrust lots of personal information to their databases – information which can then be sold, in one form or another, to advertisers and market research firms; and cloud companies often combine this information with data derived from detailed monitoring of the way you use their services. This has led one venerable Internet-watcher, Dave Winer, to liken services like Facebook to hamster cages. 'They make a wide variety of colorful and fun cages,' he writes, 'that are designed to keep the hamster, and their human owners, entertained for hours. When you get tired of one, you can buy another. It looks great until you realize one day, that you can't get out! That's the whole point of a cage.' So, he warns, 'when they say you get to use their social network for free, look for the hidden price. It's there. They're listening and watching. It's pretty and colorful and endlessly fun for you and your human owner.'[17]

This idea of Facebook as the 'owner' of its subscribers may seem paranoid, but actually it's closer to the truth than most of those subscribers seem to realize. For one thing, Facebook asserts the right to use or re-use any of the content that people post on the site. For another, it profits by not having to create any of this content – it's all done by the users.

This has led Nicholas Carr to view social-networking services as just the latest manifestation of an ancient, exploitative economic model. It's a sharecropping system, he argues, which provides the many with the tools for production but concentrates the rewards into the hands of the few (the proprietors of these cloud computing services). 'It's a sharecropping system, but the sharecroppers are generally happy because their interest lies in self-expression or socializing, not in making money, and, besides, the economic value of each of their individual contributions is trivial. It's only by aggregating those contributions on a massive scale – on a web scale – that the business becomes lucrative.'[18]

One could argue that Carr is too lenient. Classical share-cropping is a system of agriculture in which a landowner allows a tenant to use the land in return for a share of the crop produced on the land. In the online version, the virtual tenants of the social-networking space produce content that has value over and above the pleasure that it gives them and their friends; but they don't get any share of the 'crop'. So the question becomes: what is the crop in the social-networking context? Is it the pleasure and satisfaction that users receive? Or the value implicit in their personal data that the service provider is harvesting?

For the computer industry . . .

Cloud computing has serious implications for hardware manufacturers because it implies a shift in the demand for personal computers – away from the complex, sophisticated desktops and laptops of today and towards simpler devices which have less onboard storage, less powerful processors and different interfaces. This doesn't mean that the PC will disappear, just that the desktop machines of the future will be used more for tasks that require local processing power, RAM etc. rather than as life-support systems for web browsers.

For companies like Microsoft, Adobe and Apple that produce expensive PC software, the implications of the new computing topography depend on the nature of each product and the context in which it is used. When the switch to cloud services first began, it was widely believed that Microsoft Office – the dominant office suite and one of Microsoft's chief sources of revenue – would be threatened. Why would anyone pay hundreds of dollars for programs when 'free' equivalents were available over the Web? The answer, of course, is that some people won't. But then some harsh realities dawned as the capabilities of the new web services were explored.

In the first place, they turned out to be significantly underpowered compared with their standalone counterparts, and in some cases there were serious incompatibilities between web-created documents and those produced using the conventional programs. Secondly, the performance of web services depended critically on the speed and capacity of the broadband connections needed to access them. Thirdly, it became clear that the Microsoft Office suite is so embedded in the workflow of so many organizations that they would find it difficult to switch to another document-processing technology, no matter how cheap it appeared to be. And finally – and perhaps most predictably – Microsoft itself didn't stand still in the face of the threat. Instead it launched its own 'Office in the Cloud' service and moved to a position where users of Microsoft Office could move seamlessly from the desktop to the cloud and back again.

For companies like Adobe that make expensive, powerful, compute-intensive, hard-to-learn programs for creating and editing graphics, photographs and movies, the threat of cloud-based alternatives seems slight. Just as nobody in her right mind would want to revise a complex Word document on a mobile phone, nobody would want to edit a serious movie or design a building on an iPad. So it looks as though the short-term future for PC software will be determined by the ancient adage: *horses for courses*. Some things work really well on the Web, and other things still require the desktop PC. *Plus ça change.*

For mainstream businesses . . .

The effects of the Internet were initially felt in industries that were close to computing and information technology. But in the end the network has had a dramatic impact on almost every industry – not just those in the IT business. Think of bookselling, for example; or of record companies, TV networks, airlines and newspapers. We're seeing a similar process happening now with cloud computing. At present, many non-technology companies regard it as a development that has relatively little to do with them. Or if they think of it at all, they regard it simply as the latest variation on an old story, namely the 'outsourcing' of IT support functions to external contractors, including ones based overseas in countries like India. If this is what companies are indeed thinking, then they may be in for some nasty surprises.

This is because cloud computing has the potential to be very disruptive to established companies and markets. In a way, this shouldn't surprise us because it is likely to follow a trajectory that will be familiar to anyone who has followed the history of the computing industry. It starts with enabling companies to do more efficiently what they already do with their premises-based IT systems. Next it enables them to do things that they had not hitherto been able to do at all. And finally they will be able to use the IT platforms they have constructed in the cloud to move into industries that were hitherto closed to them.

'Looking ahead,' write John Seely Brown and John Hagel – two of the technology world's most celebrated sages, 'we see a series of significant disruptions that will be catalyzed by the evolution of cloud computing. These disruptions will become progressively more widespread and profound, creating opportunities not only to re-shape the technology industry but all institutional architectures and management practices in an expanding array of other industries.[19]

Hagel and Seely Brown see the evolution of cloud computing as a four-stage progression:

- Infrastructure as a Service (IaaS): the provision of 'raw utilities' such as computing power and electronic storage resources, as services over the network. This is what Amazon already offers with its S3 and EC2 services.
- Platform as a Service (PaaS): the provision of tools and environments to build and operate cloud applications and services.
- Software as a Service (SaaS). The basic idea is simple: instead of a software company selling a software licence that the customer then implements on its own data centre, the service provider hosts the software on its own computers and provides access to the system on a subscription basis. In other words, it turns software from a licence sale to a subscription service. This is already happening with organizations like Salesforce.com which provides sales support and Customer Relationship Management (CRM) software for a flat per-user, per-month fee of less than $100 with no up-front licence fee, and no software to install.[20]
- Business as a Service (BaaS). This is where a company has built a complete cloud-based system for its own purposes – for example a supply-chain management system – which it can then offer to other companies, including one operating in completely different industries. We can already see embryonic examples of this in the way that eBay and Amazon

> make their order-fulfillment and payment systems available to online traders (from whom they then take a cut). If implemented on an industrial scale, this could turn out to be the most disruptive aspect of cloud computing.

In the paper in which Hagel and Seely Brown set out this vision of the evolution of cloud computing, there is a chart showing the process as a staircase leading to the future. In reality, it's unlikely to be anything like as orderly as this. We've already noted that there is already some activity on each of the four 'steps'. And for many companies the move to cloud computing looks fraught with risks and difficulties, no matter what the potential savings are. They worry, for example, about the security of their most confidential data. If they have to rely on virtual machines which can be switched on and off at will in an 'elastic' cloud, how can they be sure that their data will have been wiped before someone else rents a virtual machine on the same physical server? 'What happens,' asks *The Economist*, when data stored on a virtual machine get compromised? Will the customer ever know? Will the provider take responsibility? That is a serious concern for companies, given the laws now in place for notifying victims of data breaches, as well as for auditing financial results for compliance reasons or e-discovery.'

The magazine goes on to report the results of a survey conducted in 2010 by *CIO Magazine*, a publication aimed at the chief information officers of large corporations. 'Although 60% of CIO's 800 or so respondents were thinking seriously about cloud computing, only 8% had committed themselves to implementing it. Meanwhile, 29% claimed to have no interest in doing so whatsoever.'[21]

The truth is that there are probably some corporate IT functions that will always be too sensitive to entrust to the cloud. But for many other, more routine functions, the economics of not having to run your own data centres will prove more powerful than reservations about the cloud. CEOs will ask their CIOs why they should pay for more computing facilities than they need when they can have the same facilities on a pay-as-you-go basis. In these cases money doesn't just talk: it positively screams.

For the environment . . .

In 2004 Apple ran an intriguing ad for its new iMac desktop computer – the one that appears to be just a screen on a stand. 'Where did the computer go?' was the caption. In fact, it was a trick question, designed to highlight Apple's design ingenuity in being able to pack the innards of an entire desktop computer into what was effectively an enclosure for a big flat-panel display. But it's a question we might also ask of the iPad, smartphone or netbook that you may now be holding in your hand. Where did the computer go?

To find the answer, come with me to a remote corner of the US – to Oregon, for example, or North Carolina. After driving for a bit from the nearest town, we come on an enormous, featureless metal shed standing on a much larger plot. It's like those 'distribution centres' you see near critical motorway junctions in the UK, only bigger. Unlike a commercial distribution centre, however, this shed is surrounded by rings of military-grade security. Nobody gets in here without due authorization.[22]

After you get past security and start your approach to the building you notice that to one side there's something that looks like an electricity sub-station. It's a transformer farm which takes high-voltage electricity from the grid and converts it into PC-friendly voltage. The only other time you've seen a transformer farm this size is next to a major electricity-generating station. On another side of the shed there are banks of squat cooling towers, with water running down the sides of what appear to be radiator grilles. And the question that comes into your mind is: this isn't anything to do with IT, surely? In fact, what it reminds you most of is 1950s-style heavy engineering.

Once inside the building, you are overwhelmed by the scale of the place. One side is a third of a mile long. Staff have to use scooters just to get around. There's a huge space full of water pipes and pumps. Here and there you can see signs of leaks – just like plumbing in the real world. Another vast space contains refrigeration gear – coolers and chillers on an industrial scale. And then you get to the heart of the plant – another vast room, this time full of cabinets containing racks of computers. The guy who runs the place tells you there are 30,000 computers in this room. And there are nine other rooms just like this.

And then you realize: *this is where the computer went.*

Cloud computing is only possible because of huge 'server farms' or data centres like the one we've just visited. Their rack-mounted PCs are the machines on which our emails, photographs, YouTube videos, documents, Facebook profiles and personal data are stored and processed. Nobody knows how many data centres there are worldwide, and the companies that run them are pretty secretive about their locations and size. But we can make intelligent guesses, because sheds this big are pretty difficult to hide. One survey reckoned that by 2008 Google alone had nineteen data centres in the US, twelve in Europe, one in Russia, one in South America and three in Asia.[23] Other companies with comparable estates of server farms include Facebook, Microsoft, Apple, Yahoo and Amazon. Data centres vary in size from the very large (190,000 square feet) to the gigantic (Microsoft has one which measures 700,000 sq. ft. and Apple is rumoured to have an even bigger one in North Carolina). So if you had an idea of cloud computing as something engagingly wispy and ethereal, think again: this is an industry with a heavy, industrial-scale environmental footprint.

But how heavy?

The environmental impact of ICT (short for Information and Communications Technology) has two main components: the resources used, and emissions produced, in the manufacture (and disposal) of equipment; and the carbon emissions resulting from the use of the kit. Both have received surprisingly little attention until recently: it was almost as if we believed that computers had 'slipped the surly bonds of earth' as the poem puts it.[24]

But they haven't. One study estimates that the total amount of fossil fuel needed to manufacture one desktop PC is 240 kilograms. 'This suggests,' the study concludes, 'that computers are much more energy-intensive than many other products we use in daily life. The fossil fuels used to produce a computer are about nine times its weight. By comparison, the amount of fossil fuels needed to produce an automobile or refrigerator is around twice their weights.'[25]

In terms of *use*, the inescapable fact is that all ICT equipment uses electricity, and that has to be generated, thereby releasing carbon dioxide into the atmosphere. In the pre-cloud era, most ICT-related carbon emissions were triggered by premises-based equipment. Given that the average power consumption of a desktop PC in the period 1980–2000 was about 100 watts, and that – at least in the industrialized world – every office worker had a desktop machine which was idle most of the time, the financial and environmental costs of ICT were staggering but went largely unremarked, so that it came as a great surprise to the world when Gartner, a leading market research company, revealed in 2007 that the computing industry is responsible for two per cent of all the CO2 released into the atmosphere – which puts ICT on a par with the aviation industry.[26] So much for slipping the surly bonds of earth.

In the years since cloud computing took off, the balance between the carbon emissions of premises-based ICT and data centres has changed. In 2007, Gartner estimated that data centres accounted for twenty-three per cent of the industry's emissions, while forty per cent came from PCs and monitors. In 2008, another study assessed the ratio in 2007 as fourteen per cent from data centres and forty-nine per cent from PCs, and predicted that by 2020 it would be eighteen per cent data centres to fifty-seven per cent PCs.[27]

2020 seems a long way ahead to be making predictions. The evolving balance between PCs and cloud devices depends on many factors. For example: early experience with the Apple iPhone suggests that there is an inverse correlation between PC use and mobile usage: owners of smartphones and Internet-enabled mobile devices tend to make less use of their PCs. So cloud computing could escalate dramatically – and PC use could correspondingly decline – over the next decade. In which case we could be looking at a future in which data centres are responsible for a quarter, a third or even a half of all ICT-related carbon emissions.[28]

On the other hand, even if cloud computing increases dramatically, there are powerful incentives for Google, Microsoft, Apple, Amazon, Yahoo *et al* to reduce their carbon footprints. One is the sheer cost of the energy required to power and cool data centres. Another incentive may come from new environmental taxes. On the other side, *disincentives* may come from public reluctance to accept such huge industrial facilities in their neighbourhoods, and the difficulty of finding locations that are close to renewable energy supplies.

What does that mean in practice?

Well, how about computing the environmental costs of one of the most common online activities – search? In 2009 a young Harvard physicist, Alex Wissner-Gross, caused a stir when he claimed that a typical Google search generates somewhere between 5 and 10g of carbon dioxide, compared with the estimated 0.02g that browsing a basic website generates for every second you view it. Why the difference? Wissner-Gross argued that it arose in part 'because Google's unique infrastructure replicates queries across multiple servers, which then compete to provide the fastest answer to your query.'[29]

Google objected to Wissner-Gross's estimate because his concept of a 'typical Google search' involved several attempts to find a desired result. The company claimed that a one-hit Google search taking less than a second produces about 0.2g of carbon dioxide.[30] Other contributors to the debate put the emissions produced by a Google search at anywhere between 1g and 10g depending on assumptions about whether users had to boot up their computer beforehand, etc. And Nicholas Carr piled into the debate by claiming that maintaining an **avatar** on the Virtual Reality site Second Life consumes 1,752kWh (kilowatt-hours) of electricity per year. 'By comparison,' he wrote, 'the average human, on a worldwide basis, consumes 2,436kWh per year. So there you have it: an avatar consumes a bit less energy than a real person, though they're in the same ballpark'.[31]

Now I know what you're thinking: this is the modern equivalent of those medieval debates about how many angels could dance on the head of a pin. But there is a difference: those angels generated no carbon dioxide. Cloud computing, on the other hand, definitely does.

And over the horizon?

This is where cloud computing looks really interesting, because essentially it lowers the barriers to entry for innovators. Here are some stories that illustrate the possibilities:

1. Imagine a start-up with a good idea for a new kind of web service. Its founders have assembled a great team of software developers who have created an innovative product. An initial beta-release to invited users has generated intense interest. But the founders know that if they go public without the right IT infrastructure then their servers will be overwhelmed, and they run the risk of being discredited before they get started simply because their systems couldn't cope.

   In the old days they'd have to persuade investors to fund a big enough network of servers to meet the anticipated demand. Now all they need to do is sign up to use Amazon's S3 and EC2 services. If the anticipated demand does indeed materialize, then the company can rent more virtual machines, and keep ramping up the provision until the demand has been satisfied. And the only outlay the start-up has to make is the rent for the servers and the disk storage that it has used.

If this seems far-fetched to you, then let me name a start-up which faced exactly this challenge – and solved it in the way I've described. The company is Twitter.

2. Or take another example – a company that develops large, complex software products. Most of the time it doesn't need much in the way of computing resources, beyond what its team of developers needs. But there comes a time in the development of the product when really intensive testing has to be done to flush out bugs and potential problems with the software. All of a sudden the company needs thousands of test machines. Once upon a time, that would have posed a major headache. Now it's a problem with a simple solution: cloud computing services.
3. A third example: an important national newspaper has made a major strategic decision. It's going to scan its entire archive and make it available online as PDFs[32]. To do that it has to translate every scanned page image into that format. The engineer in charge of the project estimates that conversion to PDF will take a year on the newspaper's own in-house computing system. So he uses his personal credit card to book virtual machines on Amazon's Elastic Computing Cloud, uploads the data – and the entire job is completed in 24 hours.
4. A final example: a start-up company which has been set up to develop and commercialize a new discovery in genetics. Its founders are the scientists who made the original discovery. But getting the product ready for market requires an enormous amount of computing to explore possible molecular structures. In the old days, the only option would

> have been for the team to seek access to one of the supercomputer centres maintained by national governments – access that might or might not have been granted. But now all the computing power they could conceivably need is available simply on production of a credit card.

It's stories like these that have led some enthusiasts to speculate about whether cloud computing will create the conditions for a contemporary version of a 'Cambrian explosion' – that geological period 530 million years ago in which the speed of evolution seemed to accelerate and the number and diversity of species grew to resemble what we have today. This seems a bit hyperbolic. What does seem probable, though, is that the world envisaged by John Gage in the 1980s is going to prove disruptive. Many innovations – in business as well as in IT, science and medicine – require temporary or episodic access to massive computing resources. Before the advent of cloud computing, these resources were pretty hard to come by. Now they're affordable – and running on the network.

7. THE WEB IS EVOLVING

In Chapter 2, I made the point that the Web isn't the Internet – any more than Google or Facebook are – but simply an application that runs on the network's infrastructure. The Web is important, but it's not the only game in town. If you monitor data traffic on the network then you will find that a significant proportion is taken up with web activity. But you'll also find significant levels of traffic generated by other applications – email, Internet telephony, video, file-sharing, file-transfers, streaming media (like the BBC iPlayer), to name just a few. Nevertheless, for many people, the Web is their main point of contact with the Internet – which may help to explain why they tend to confuse the two.

Fundamentals

So what exactly is the Web?

Here are two definitions:

A computer network consisting of a collection of Internet sites that offer text, graphics, sound and animation resources through the hypertext transfer protocol.[1]

The set of all information accessible using computers and networking, each unit of information identified by a URI.[2]

Not terribly helpful, are they? For most of us, a working definition of the Web would say something like this: the World Wide Web is *a vast store of online information in the form of 'pages' of varying sizes which are connected to one another via clickable links*. But the moment you start to interrogate this common-sense definition, it starts to come unstuck.

For example, it implies that the basic elements are static 'pages' which are stored on a computer somewhere on the Internet and 'served' to anyone who clicks on the link corresponding to the page's address. But many such pages don't exist *until you click on the link*. Their contents reside in various databases and the page is assembled on the fly by the server before it is dispatched to you. So immediately your mental model of the Web as a vast trove of static documents is called into question.

The definition also implies that all you do with the page is passively to view it. But in fact it is just as likely that you will *interact* with it, and that it will respond to your actions. If you go to Ryanair.com to book a flight, for example, the site sends you what appear to be web pages, but each of them is effectively a small virtual computer which can respond to your inputs. That's why, for example, if you try to book a flight without confirming that you have read and understood Ryanair's terms and conditions, the page will pop up a dialog box telling you that you cannot proceed until you have ticked the relevant box. How is this done? Does Ryanair check to see whether you've ticked the box before sending you the flight reservation page? Not a bit of it: within the web page there's a small program which runs in your browser and does the checking without ever troubling Ryanair.com.

THE WEB IS EVOLVING

There was a time when the Web was indeed a trove of static pages. But those days are long gone. The gap between our simplistic mental models and the reality of contemporary e-commerce sites is a reflection of the extent to which the Web has changed since its inception in 1990–1.

And it's still changing. In fact, we can look at its evolution in much the same way that earth scientists look at the history of our planet – as a succession of geological eras. But whereas geologists have fancy Latin names for their eras – *Protozeroic*, *Paleozoic*, *Mesozoic*, and so on – engineers prefer simpler nomenclature based on the conventions for labelling software releases.

In those terms, the Web has, to date, been through two major geological eras – Web 1.0 and Web 2.0 – and is now slowly moving towards a third – Web 3.0. And for each era, the answer to the question 'What is the Web?' is different, which perhaps helps to explain why our assumptions about the system are sometimes out of sync with the reality.

We will look at these Web 'eras' in more detail in a moment, but first let's try and get a handle on the essence of the thing. In engineering terms, the Web has *three* key components:

- A set of digital *resources* which are stored on Internet-connected computers across the world.
- *Protocols* which provide: (i) a standard way of assigning a unique, machine-readable address to each resource; (ii) a mark-up language for structuring the resource in a way that enables it to be accurately rendered by browsers; and (iii) conventions for conducting interactions between computers concerning the transfer of web pages.[3]
- *Software* for serving resources on demand, and for rendering them when they arrive at their destinations.

These were the components that Tim Berners-Lee and his CERN colleagues envisaged and programmed in the astonishing burst of creative energy in 1989–90 in which the Web was conceived. And they are still at the core of the system, though they are now surrounded by a thicket of technologies that have evolved as the Web has grown and diversified.

Vital statistics

How big is the Web? It's impossible to give a definitive answer, for a number of reasons. Firstly, there's the issue of static versus dynamic (i.e. rendered-on-the-fly) pages. In 2008, Google claimed that its software had identified one trillion unique URLs. In fact, the company's engineers reported that they had found even more than a trillion individual links. But not all of them led to unique web pages. 'Many pages,' they explained,

> have multiple URLs with exactly the same content or URLs that are auto-generated copies of each other. Even after removing those exact duplicates, we saw a trillion unique URLs, and the number of individual web pages out there is growing by several billion pages per day.
>
> So how many unique pages does the web really contain? We don't know; we don't have time to look at them all! :-) Strictly speaking, the number of pages out there is infinite – for example, web calendars may have a 'next day' link, and we could follow that link forever, each time finding a 'new' page. We're not doing that, obviously, since there would be little benefit to you. But this example shows that the size of the web really depends on your definition of what's a useful page, and there is no exact answer.[4]

Secondly, and more importantly, there's the fact that the Web is like an iceberg in that the visible part of it – i.e. the part that is accessible to search engines – may well be the smallest bit. What's more, we don't know the relative proportions of the visible and invisible parts because the 'darkweb' – the part that search engines cannot reach – is made up of resources that are locked away behind corporate firewalls or media 'paywalls', and so its size has to be a matter for conjecture rather than measurement.

As far as the visible part is concerned, as I write this the most frequently cited measurements put its size at between twenty and forty billion pages.[5] You may say that there's a hell of a difference between twenty billion and forty billion, and there is. But given such overall magnitudes, the margin of error pales into insignificance. The only thing one really needs to know is that the Web is unimaginably vast – which is why it was reasonable to draw a comparison between Tim Berners-Lee and Gutenberg at the beginning of this book, for both men's inventions had an impact on the world that is beyond computation.

The Web's geological eras

The reason many Internet users are unaware of how the Web has changed over the years is the same reason why your children's grandparents are more conscious than you are of how your kids have changed: they see them less frequently. Living with them on a daily basis you are less likely to see the small changes that cumulatively amount to radical transformations.

So it is with the Web, which is why it is intriguing sometimes to check out the Internet Archive, a not-for-profit organization which captures snapshots of the Internet over time.[6] The Archive runs a fascinating service called the Wayback Machine, which enables one to search for old web pages. For example, Fig 7.1 shows how the home page of Yahoo! looked in 1996. And Fig 7.2 shows what it looked like fourteen years later.

In 1996, Yahoo's home page was essentially just a list of hyperlinks. In 2010, the same site offers webmail, interactive gaming, instant messaging and video – all apparently within the confines of the home page. What we're looking at is the difference between two geological eras. Yahoo in 1996 was typical of Web 1.0; in 2010 it has elements of something different which some people call Web 2.0. So let's now look at what those terms mean, and how the transition from one to the other came about.

Web 1.0

The Web was originally conceived as a way of sharing information among particle physicists who were scattered across the world. Since most of that information was in the form of documents, the design was therefore for a system that would make it possible to format these documents in a standardized way, publish them online and make them easy to access. So the first 'release' of the Web (to use a software term) created a world-wide repository of linked, static documents held on servers distributed across the Internet.[7]

The original design was based on two key concepts. The first was the notion of *hypertext* – i.e. an older technology that allowed a computer user to click on a spot in a document that would then take them to a defined place in another document.[8]

The second underlying concept was that the Web should

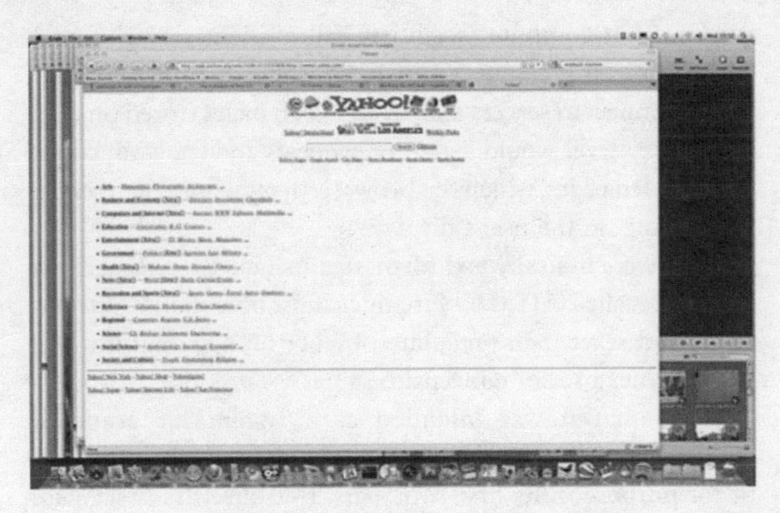

Figure 7.1: The Yahoo! Home page in 1996

Figure 7.2: The Yahoo! Home page in 2010.

run on what computer scientists call a 'client-server' model. The *client* in this case was a program called a 'browser' which would connect to servers and request web pages stored on their disks. The server would listen for requests and transmit copies of requested pages, which the browser-client would then render for viewing on the user's display.

Pages were basically text formatted using a special mark-up language called HTML[9] Finally, communications between client and server were regulated by HTTP, a special Internet *protocol* (i.e. a set of conventions).[10]

Given that it was intended as a system for academic researchers, the original Web design was probably admirably fit for purpose in its first two years. But once the first major graphics-capable browser – Mosaic – appeared in the spring of 1993 and the commercial possibilities of the technology became obvious to the corporate world, the limitations of the original concept began to grate. The early Web didn't make provisions for images, for example. And it was a one-way, read-only medium with no mechanism for enabling people to interact with web pages – which meant that it was unsuitable for e-commerce. There was no way for users to talk back to authors or publishers; no way to change or personalize web pages; no way to find other readers of the same page; and no way to share or collaborate over the Web.

From 1993 onwards, therefore, there was a steady accretion of innovative technologies designed to extend Berners-Lee's creation and overcome some of its perceived limitations. The main driver behind this was e-commerce, which desperately needed to transform the Web into a medium that facilitated *transactions*.

In order to make transactions possible, a whole range of

problems had to be solved. For example: ways had to be found to allow interactivity between browsers and servers, to facilitate personalization of web content and to overcome the problem that the HTTP protocol was both insecure (in that communications between browser and server could be intercepted and monitored by third parties) and stateless (i.e. unable to support multi-step transactions).

In time, solutions to these problems emerged. Personalization was achieved by using *cookies* i.e. small, hidden text files which websites deposited on users' computers, thereby enabling a site's operators to detect whether a visitor was a new arrival or a repeat customer, what their preferences were, and so on. Insecurity was addressed by the development of HTTPS, a secure version of the HTTP protocol that encrypted all communications between browser and server. Multi-stage transactions ('sessions') were enabled by having the browser engage with *scripts* (small programs) running on the server. And browsers themselves evolved to cope with the additional functions that they were increasingly called upon to perform – often by having capabilities added by specialized add-on programs (called *plug-ins*) like Adobe's Flash which enabled them to handle audio and video and other kinds of file.

Many of these technologies which drove the evolution of the Web from 1993 onwards had an ad hoc feel to them. They were what in Britain would be called lash-ups. (Indeed the PERL programming language, one of the original tools for building them, has been affectionately described by its devotees as 'the duct tape of the Web'.)[11] This was hardly surprising, given that they had been grafted onto a system rather than being designed into it. But these improvisatory technologies proved extraordinarily powerful in supporting the dramatic expansion of the Web from 1995 onwards. Some of the lash-ups' technical limitations seemed onerous almost from the beginning, however. For example, supporting transactions by requiring intensive back-and-forth communications between a browser running on a consumer's desktop and a script running on a distant server seemed like a clumsy way of using bandwidth, not to mention users' time. Wouldn't it be more efficient to have some of this computation done in the browser?

The answer, of course, was yes, and in late 1995 it emerged. It was called Javascript – a programming language that enabled site operators to embed small programs in web pages that can be run within the browser that is displaying them. For example, the page can pop up a window in response to something that the user does; or it can check the validity of what is typed into a web form – so that if you accidentally specify a return date for a flight that is earlier than the departure date, then the web page can alert you to the error; or it can change images on the page as your mouse rolls over them. And all of these things can be done without the distant server having to do anything.

From 1995 onwards, the Web entered a period of explosive growth – mainly triggered by the wildly successful stock market flotation of Netscape (the first company to market a web browser and server software) in August of that year. This persuaded Wall Street (and investors around the world) that there was serious money to be made from investing in so-called 'dot-coms', i.e. start-up companies with business plans based around the Web.[12] The irrational exuberance implicit in this seemed obvious to many detached observers almost from the outset, but as always happens with speculative bubbles, the thing had to run its course. Which it duly did until, in 2001, the bubble burst. And suddenly, with the 20/20 vision of hindsight, the stock market concluded that the Web was not the future, but the past.

Web 2.0

Geological eras are inevitably assigned names retrospectively, since they predate by millions of years the humans who thought up the nomenclature. With Web eras, it's different, because those doing the naming have lived – and are living – through the upheavals they aspire to label. What happened with the Web was that it evolved from Web 1.0 at a furious pace under the combined pressures of commerce, software development, entrepreneurship, increasing bandwidth and a vast growth in the Internet-using population. And then sometime in 2004 a few perceptive observers began to ponder the significance of what had been going on. There was a certain urgency to their deliberations because they needed a title for a conference they were organizing on how the Web was changing.[13] In the end, they plumped for something rather prosaic: they called it Web 2.0.

The conference in question was staged by a company called O'Reilly Media (formerly O'Reilly & Associates), a California-based firm that is the world's pre-eminent publisher of books on technical aspects of computing.[14] It was founded by Tim O'Reilly and its success has been largely due to his skill at spotting technological trends, inferring from them the specialist knowledge that will be required to service them, and producing publications that meet that anticipated demand.[15]

In pondering the collapse of the first Internet bubble, O'Reilly drew a different conclusion from most other observers. Whereas they saw the bust as evidence that the Web had little real economic substance, he saw it as a salutary sorting of technological wheat from speculative chaff. 'The bursting of the dot-com bubble in the fall of 2001,' he wrote later, 'marked a turning point.'

> Many people concluded that the web was overhyped, when in fact bubbles and consequent shakeouts appear to be a common feature of all technological revolutions. Shakeouts typically mark the point at which an ascendant technology is ready to take its place at center stage. The pretenders are given the bum's rush, the real success stories show their strength, and there begins to be an understanding of what separates one from the other.[16]

In pondering the Web 1.0 enterprises that had survived the crash, and the new ones which had arisen after it, O'Reilly and his close colleagues began to realize that they had several important features in common.

1: Harnessing collective intelligence

The first, and most important, characteristic is that they had all – in one way or another – 'embraced the power of the Web to harness collective intelligence'. In some cases – as with Google's PageRank algorithm (which ranks web pages on the basis of how many other pages link to them – a kind of automated peer-review) – this was done via software analysing the hyperlinking that is at the core of the Web. In other cases (eBay, Amazon) it was by exploiting the willingness of users to engage with the enterprise. O'Reilly pointed out that Amazon sells the same products as competitors like Barnesandnoble.com and receives the same product descriptions, cover images and editorial content from publishers. But:

> Amazon has made a science of user engagement. They have an order of magnitude more user reviews, invitations to participate in varied ways on virtually every page – and even more importantly, they use user activity to produce better search results. While a Barnesandnoble.com search is likely to lead with the company's own products, or sponsored results, Amazon always leads with 'most popular', a real-time computation based not only on sales but other factors that Amazon insiders call the 'flow' around products. With an order of magnitude more user participation, it's no surprise that Amazon's sales also outpace competitors.[17]

Other examples of harnessing collective intelligence were Wikipedia, and social-bookmarking sites like delicious.com which enable people to share URLs of web pages and sites that they find useful or interesting. Wikipedia was made possible by the invention, by an American programmer, Ward Cunningham, of the *wiki* – a web page that could be edited by anyone who read it in 1995, and in the process transformed the Web from a one-way, read-only medium into what Tim Berners-Lee later called the 'read-write Web'.

2: User-added value

A second distinguishing feature of the 'new' Web was what came to be known as 'user-generated content' or 'peer production' – that is, material created and published freely by people who do it for no apparent economic motive.[18] The prime example was blogging, but others included services like Flickr, Smugmug and Photobucket which freely hosted photographs uploaded by their subscribers, and (later) Vimeo, Blip.tv and

YouTube which did the same for video material.

Finally, among the new arrivals on the Web after the dot-com crash were social-networking sites (MySpace, Orkut, LinkedIn, Facebook, for example) which were explicitly designed to build (and then exploit) new kinds of online communities by harnessing the incurable sociability of human beings.

Implicit in all these enterprises, O'Reilly realized, was a simple proposition: *users add value*. 'But,' he noted, 'only a small percentage of users will go to the trouble of adding value to your application via explicit means'. Therefore, the new enterprises set 'inclusive defaults for aggregating user data and building value as a side-effect of ordinary use of the application. . . . They build *systems that get better the more people use them*.'[19]

There was, of course, another side to all this, in the sense that the business model behind many of the sites hosting user-generated content was brilliantly simple: users did all the work – created the content, added the metadata and uploaded it to the site, all without payment – while the site owner reaped the rewards by converting web traffic and aggregating user data into advertising revenue. Facebook, Flickr and the others provided the space on the Web; users did the rest. As we noted earlier, this is what led one astringent critic, Nicholas Carr, to liken the business model of these services to a much older one – sharecropping. 'Web 2.0,' he wrote, 'provides an incredibly efficient mechanism to harvest the economic value of the free labor provided by the very many and concentrate it into the hands of the very few.'[20]

3: Hidden wiring: RSS and APIs

Another distinctive feature of the 'new' Web was that many of the emerging services on it were dynamically interconnected by means of software.

One important technology for doing this had emerged towards the end of the first Internet boom, in 1999. It was known as RSS. After some initial disputes about whether it should be called 'RDF Site Syndication' or 'Rich Site Syndication' it was eventually rechristened 'Really Simple Syndication.' Its significance is that it provided a way for a website automatically to generate a 'feed' – i.e. a summary of recent updates which could be subscribed to by other sites.[21] Among other things, this meant that the web browser was no longer the only way of viewing a web page. Another kind of program – the **feed reader** – emerged which could give one a different way of keeping up with what was being published on the sites to which one subscribed. It was more akin to the stream of information that comes from a newspaper wire service – terse, timely, always up-to-date – and, most importantly, machine-readable.

The other technology used for facilitating interconnectedness harked back to the early days of PC software and was based on the idea of an Application Programming Interface or **API**. Just as a Human-Computer Interface (HCI) like Windows enables people to interact with computers, an API is a set of technical conventions that enables one program to work with another. Microsoft publishes APIs for its *Windows* operating system, for example, which enable programmers to write specialized software that will run under that operating system. APIs provide the 'hooks' onto which other pieces of software can hang. But until the late 1990s, these were mainly used to specify interactions between software running on specific *machines*. What was distinctive about some of the web services that evolved after 1999 was that they used APIs to specify how *entire web services* could work together.

A typical example is the API published by Google for its

Google Maps service. This made it possible for people to create other services – called 'mashups' – which linked Google Maps with other Internet-accessible data sources. A well-known example was housingmaps.com, which combined Google Maps with apartment rental and home purchase data from the Craigslist classified advertising site to create an interactive housing search tool.

RSS and APIs provided the hidden wiring for the new Web that emerged in the early years of the millennium, transforming it into what David Weinberger articulated as 'small pieces, loosely joined'.[22] Many people missed the significance of this observation because they assumed that the dense interaction between services implied the emergence of a unitary system along more traditional lines where, over time, the various components would be coordinated. But this was to misunderstand what was happening. As Tim O'Reilly put it, 'Simple web services, like RSS . . . are about syndicating data outwards, not controlling what happens when it gets to the other end of the connection. This idea is fundamental to the Internet itself, a reflection of what is known as the end-to-end principle.'[23] Once Google has published the API for Google Maps, it has no responsibility for what people do with it. Nor does it provide any guarantees about service levels or continuity. Which of course means that anyone building a business on the back of an API may be taking a risk.

4: Perpetual beta, continuous change

This intrinsic, experimental ethos of the emerging Web was exemplified by one of its poster children, the Google search engine. When it launched, and for a considerable time afterwards, the Google logo carried the subscript 'BETA'.

The term comes from the terminology used by the computing industry to describe stages in the software development and release cycle.[24] The first stage, sometimes called *pre-alpha*, is when the software is being written. Then comes the *alpha* phase, when the program is largely complete and ready to be subjected to initial testing. The alpha version is always assumed to have bugs and to be unstable – i.e. liable to cause system crashes which may involve loss of data. The developers may also feel free to add or remove features from the program. In short, the alpha version is really only for knowledgeable users who are willing to try it at their own risk. The end of this phase is usually marked by *feature freeze*, after which no more major additions are made to the program.

The *beta* version is the one deemed suitable for testing by a subset of its target market because it is stable and less likely to cause system crashes. So a *beta release* will go to hundreds or perhaps thousands of existing or prospective customers, together with provisions for reporting their experiences with the software. Some beta releases are completely open – in the sense that a copy of the program may freely be downloaded from a website. Closed or private betas, in contrast, are made available only to invitees. In either case, though, the objective is the same: to expose the software to as wide a range of use as it is likely to get in order to unearth bugs, unanticipated interactions and other deficiencies.

The final stage in the software development cycle comes after beta testing and is the first release of the product, usually labelled 'Version 1.0'. Subsequent minor upgrades (bug fixes etc.) are signified by incrementing the number after the decimal point – v1.1, 1.2 and so on. Major upgrades are signified by incrementing the main version number – v2.0, 3.0 etc.

Figure 7.3: An early version of the Google logo.

Historically, this nomenclature evolved in the context of packaged software designed for mainframes, minicomputers and PCs. It came from a world in which people bought software in shrink-wrapped boxes and installed it on their own computers. It was also a world in which software upgrades rippled slowly and painfully through their established customer base. People had to be persuaded to purchase and install the newer versions. In the process, they often discovered that their PCs also needed upgrading simply to be capable of running the new versions. And organizations sometimes blanched at the cost and disruption involved in upgrading, say, from one version of Windows to another.

This Byzantine software-upgrade process had one formidable downside: it acted as a sheet-anchor on innovation because software companies had to strike a balance between two competing obligations. On the one hand, there was the desire to create more powerful and feature-rich programs and to keep up with the competition. On the other hand, there was the obligation not to annoy one's existing customers by ensuring that the new software would still run on so-called 'legacy' systems: older machines, older peripherals (like printers) and earlier versions of operating systems. And the irony was that the more successful a software company became, the worse this problem of backward-compatibility became.

What was significant about the 'BETA' in the early Google logo was that it signalled its designers' determination to cut loose from the sheet-anchor of legacy systems. From the outset, they regarded their web-based service as *work in progress* – subject to continual and sometimes rapid change – rather than as something fixed and immutable. What made this possible, of course, was the fact that it was a cloud-based service, so every user's version of the software could be upgraded at a stroke, and without any effort on their part – beyond occasionally upgrading their browser software or installing some (free) plug-ins designed to take advantage of whatever new features Google had decided to add.

This philosophy of 'perpetual beta' infected many of the Web enterprises that appeared from 1999 onwards, and was at least partly responsible for the creative explosion that they embodied. This was because implicit in the idea of a perpetual beta is the fact that the service got better the more people used it. In the Open Source programming community there is a saying that 'given enough eyeballs, all bugs are shallow' and it applied with a vengeance to the new Web-based services with their millions of daily users. Deficiencies and bugs were remorselessly exposed – and also quickly fixed, without requiring those millions of users to upgrade their own software. O'Reilly quotes a web developer at a major online service saying: 'We put up two or three new features on some part of the site every day, and if users don't adopt them, we take them down. If they like them, we roll them out to the entire site.' And he reports a revelation by the lead developer of the photo-sharing service Flickr that they deployed new builds[25] up to every half hour. 'This is clearly,' observes O'Reilly, 'a radically different development model! While not all web applications are developed in as extreme a style as Flickr, almost all web applications have a development cycle that is radically unlike anything from the PC or client-server era.'[26]

This is why the emerging Web environment marked such a radical departure from the world where the PC was the computer and in which Microsoft was the dominant software supplier. Or, as one commentator put it, 'Microsoft's business model depends on everyone upgrading their computing environment every two to three years. Google's depends on everyone exploring what's new in their computing environment every day.'[27]

5: The Web as a platform

A final distinguishing characteristic of the post-1999 Web was that the enterprises and services that were becoming dominant were effectively using the Web as a *programming platform*. So while the Internet was the platform on which Web 1.0 was built, Web 1.0 became the platform on which the services characteristic of Web 2.0 were constructed.

The perils of lexicography

No sooner had O'Reilly *et al* attached the label 'Web 2.0' to the developments we have been reviewing than they were attacked from all sides. It was said (correctly) that they were not the first people to have coined the term. Tim Berners-Lee pooh-poohed it, saying that the label was just another name for 'the read-write Web'. Others criticized it for lacking the precision of a proper definition: it was, they said, just a vague phrase which failed to provide unambiguous criteria for determining

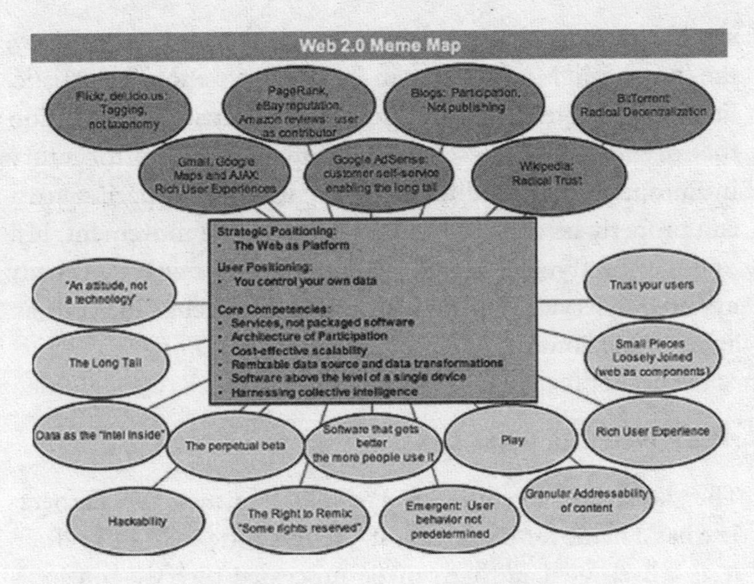

Figure 7.4: A Meme Map of Web.2.0. From Tim O'Reilly, 'What is Web 2.0?' http://oreilly.com/pub/a/web2/archive/what-is-web-20.html?page=1

what should count as a genuine Web 2.0 service. In response, O'Reilly conceded the point. 'Like many important concepts,' he wrote, 'Web 2.0 doesn't have a hard boundary, but rather, a gravitational core. You can visualize Web 2.0 as a set of principles and practices that tie together a veritable solar system of sites that demonstrate some or all of those principles, at a varying distance from that core.' He illustrated the point with what he called a 'meme map'.[28]

In the end, the quest for definitional precision seems futile to me. Apart altogether from the fact that the Web is changing so fast that it will always be a moving target for lexicographers, a label like Web 2.0 is really just a way of describing a significant technological shift in much the same way as, say, the 'Romantic Movement' can be used to denote an artistic, literary, and intellectual reaction to the Industrial Revolution that originated in the second half of the eighteenth century in Europe. It may be difficult to say whether a specific artist can properly be regarded as belonging to the movement, just as it may be arguable about whether a specific website is truly a Web 2.0 service, but it's still useful in placing them in an appropriate context.

After Web 2.0, what next?

The glib answer, of course, is 'Web 3.0'. As usual, Tim Berners-Lee has a name for it. He calls it 'the semantic web' and defines it as 'a web of data that can be processed directly and indirectly by machines.'[29] It builds on the Web 2.0 idea of interacting services whose connections are mediated by software and APIs and adds to that ways of enabling computers to make more informed interpretations of the contents of web pages.

To see what's behind the idea, you need to recall the difference between syntax and semantics. Syntax is about grammatical rules – how words are arranged into sentences. Semantics is about what those sentences mean.

Consider the sentence 'I love New York'. If I delete the word 'love' and substitute instead a heart symbol, then I will have changed the syntax, but the semantics of the sentence remain unchanged. As a human being, you have no difficulty in understanding that. But to a computer, the change would be completely baffling. Computers (or, more accurately, computer software) can *parse* text – i.e. analyse its structure to determine its grammatical structure; but they have great difficulty in divining its meaning. In other words, semantics is beyond them.

Why is this a problem? Well, simply because it means we cannot unlock the full potential of the Web. There are all those billions of web pages out there, packed with useful – or potentially useful – data, but for the most part the only way we can convert those data into meaningful information or knowledge is by getting human intelligences to process them. What we'd *really* like to be able to do is to enable machines – computers – to 'read' the data on web pages, determine what they mean and do intelligent things with the resulting information. Such a system, for example, would be able to give an intelligent response to a question like: 'I'm looking for somewhere to go on holiday that is reasonably hot, has good food, is within two hours' flying time, and costs no more than £300 per person per week. Oh – and I'm a vegetarian.'

The goal behind the semantic web is attractively simple. It is to add 'a layer of meaning on top of the existing Web that would make it less of a catalog and more of a guide – and even provide the foundation for systems that can reason in a human fashion'.[30]

The goal might be simple but achieving it will be formidably difficult, largely because most web pages are couched in natural language and computers are hopeless at dealing with that. Humans have no difficulty in coping with ambiguity, for example, so that we can easily distinguish the different meanings implied in the pair of sentences:

- 'Time flies like an arrow'
- 'Fruit flies like a banana'

whereas a machine would be totally confused by them.

There are engineering solutions to the problem of getting computers to understand the semantics of web pages, but they involve imposing fairly elaborate structures on the pages themselves. The World Wide Web Consortium – the body that sets technical standards for the Web – has proposed a way of doing the structuring based on two elements: code-named RDF and OWL. RDF stands for *Resource Description Framework* and provides a stilted way of describing elements of a web page. Thus in RDF the sentence 'The sky has the colour blue' is expressed as a 'triple' of specially formatted text strings where:

- 'Sky' is the *subject*;
- 'has the colour' is the *predicate*; and
- 'blue' is an *object*.

OWL stands (confusingly) for *Web Ontology Languages* and is intended to provide machine-readable descriptions of facts – e.g. the fact that 'DOG is an ANIMAL' or 'DOG has four LEGS'.

Now, dear reader, if your eyes are beginning to glaze over – or if you are beginning to think that if this is what we have to do to web pages to make them comprehensible by machines then computers must be even more stupid than we had hitherto imagined – I am with you all the way. But at the moment this is the kind of structuring that experts think needs to be imposed on web pages to make the goal of the semantic web a reality. And to make matters worse, for most of the existing Web, this would have to be done retrospectively. One doesn't have to be a rocket scientist to guess that it's unlikely to happen.

So it looks as though most of the 'old' Web will remain decidedly *un*semantic. The outlook is different for web pages that are created from now on, though, because the tools we use for composing them can impose the RDF/OWL structuring on them. This is particularly true for web pages that predominantly contain empirical data. Most organizations these days collect and curate large quantities of statistical data. Increasingly, these data are published on the Web, and there is now a powerful movement aimed at persuading public authorities in particular to publish their data in ways that can enable them to be read and interpreted not just by humans but also by computers. The US and UK governments have, in the last two years, begun to publish huge volumes of data in this form, leading enthusiasts to argue that we are moving from a *Web of documents* to a *Web of linked data*.[31]

What's the point of doing this? Simple: there's lots of useful knowledge buried in those statistics, and if we can only get at it and apply computerized analytical techniques to it then we may learn things which may be useful in either making our public authorities more efficient or making them more accountable. So, for example, we can take official crime statistics on a street-by-street level and map them onto Google Maps to highlight relative differences in crime levels between different neighbourhoods. This may be disturbing for estate agents, but it could also be useful to police authorities monitoring how their local force allocates its resources.[32]

Other mashups have been created by taking UK Department for Transport data of all the road accidents in 2007 involving cyclists where injuries were reported to the police. There were just over 16,000 accidents in total – about 46 per day. 'Thanks to the raw data released by Direct Gov,' the mashup's creators write,

> we were able to learn the grid reference for each . . . We turned these into latitude/longitude coordinates using a program one of our coders wrote, which we then plotted on a map. The result is a richly detailed – and hopefully educational – portrait of the nation's cycling accidents. Zoom in and you can literally see the placemarks for individual accidents: in Richmond Park, in London, there were 8, all around the main entrances. The Isle of Skye escaped without any. The majority, as you might guess, were concentrated in densely populated urban areas.[33]

You can see why the possibilities of this are exciting. It could enable policy-makers, for example, to see whether there were fewer accidents on cycle paths, or to what extent accidents are more likely to happen at street corners than at other places, or whether towns with fewer accidents simply have fewer cyclists – or whether (as many cyclists suspect) so-called 'traffic-calming' measures actually make roads less safe for cyclists.

These are the kinds of possibilities that excite the inventor of the Web. In a memorable interview[34], he used the analogy of what PC software enables us to do with the data in our bank statements when we download them into a spreadsheet program like Microsoft Excel: 'We summarise them, we graph them, we see how we're doing from year to year, we pull out taxes from them. We do all kinds of things – create lots of different views, and we have software that is very specifically written to allow us to do the sort of things we expect to do with financial data.'

The challenge of putting data on the Web, he says, is to enable us to do the same kind of thing, but on a global scale:

> What we are looking for from the web of data is, in a way, what we found from the web of documents – a sea change. When we started publishing documents on the Web we were just publishing documents so you could go and read them. But because a hypertext link could go to absolutely any document on the Web, independent of who [its author] works for, or where they live, or what language they speak – because that hypertext link was so powerful there was a sea change in the way businesses worked. There were sea changes, too, in how people ran their lives, how they booked flights, how they ran their families. There were sea changes in how people kept in touch with each other. Things which just could not be done before became possible.

With the web of data, Berners-Lee thinks that the sea change may be even greater,

> because not only are we enabling people to access stuff more effectively across the globe; they'll be able to get at stuff they haven't been able to get at before. And because this is data that can be manipulated we can put machine power into filtering it and allowing [people] to join it to other summary data to maximize the human ability to see the correlations or the trends.

This means, he says, that mankind will be able to solve some problems that have hitherto lain beyond our grasp.

At this stage, it's impossible to say what the scale and scope of Web 3.0 will be. The only things that are clear are, firstly, the direction of travel – which is towards a machine-readable Web, and, secondly, the philosophy of those who, like Tim Berners-Lee, want to see the Web evolve in this direction. They are convinced that, in its present form, the Web is just a shadow of what it could be. In their view, Web 2.0 was an advance on Web 1.0, but it's still less than the sum of its parts. The dream behind the move to Web 3.0 and beyond is finally to make the Web deliver on its potential to become the power-steering for our collective intelligence.

8. COPYRIGHTS AND 'COPYWRONGS' OR WHY OUR INTELLECTUAL PROPERTY REGIME NO LONGER MAKES SENSE

We live by our wits. Of course we also need physical resources – natural and capital – but fundamentally our economies run on ideas, or the things that ideas have made possible. Once upon a time, physical resources were probably the deciding factor in determining whether a society thrived or declined, but ever since the Industrial Revolution knowledge has become increasingly important in determining whether we are prosperous or not.

And now we live in a knowledge economy.

Copying

Copying is at the core of this economy. Humans are – well most of us are, anyway – *copycats*. When I say this in lectures to business audiences, I can see them bristle with indignation. Then I ask a simple question: 'Please raise your hand if you have *never, ever* borrowed an idea from someone else's PowerPoint presentation'. Very few people raise their hands.

The entire vast global fashion industry runs on copying. So

do the movie, publishing and technology industries: someone makes a big hit with a new type of film (*Avatar*, say), or an intriguing kind of book (the Harry Potter stories), or a new kind of computing device (netbooks, tablets) and shortly afterwards the market is overrun with similar kinds of product. The same is true of the automobile industry, which is full of fiercely competitive companies which nevertheless magically seem to converge on similar shapes for headlight clusters.

There's nothing wrong with this. *Au contraire*, it's perfectly natural.[1] It happens because ideas have two interesting properties. In the first place, they're *infectious*, which happens because we are social animals, and social animals are imitative: that's how original ideas get disseminated in the copying process. But ideas are also *promiscuous*: that is to say, they like mating with one another, as Matt Ridley's metaphor puts it.[2] And when they do, they produce offspring which have some of the characteristics of their parents, and perhaps also mutations which are unpredictable and startlingly original.

But in order to be free to mate, ideas have to be free agents: they have to be allowed to interact with one another. So a creative society needs lots of unattached ideas and an environment in which they are able to encounter others. That implies that much of what we call creativity – the generation of new ideas – is social. This doesn't mean that some great flashes of originality don't emerge from a *Eureka!* moment; they do – sometimes. But mostly, as Steven Johnson argues,[3] creativity is a social process. 'We are often better served,' he writes,

> by *connecting* ideas than we are by protecting them. Like the free market itself, the case for restricting the flow of innovation has long been buttressed by appeals to the 'natural' order

of things. But the truth is, when one looks at innovation in nature and in culture, environments that build walls around good ideas tend to be less innovative in the long run than more open-ended environments. Good ideas may not want to be free, but they do want to connect, fuse, recombine. They want to reinvent themselves by crossing conceptual boundaries. They want to complete each other as much as they want to compete.[4]

But this raises a dilemma. On the one hand, a creative society needs a dependable flow of free ideas in order that this process of connecting, fusing and recombining can go on. On the other hand, creative people can't live on thin air. If they're going to be enabled – let alone encouraged – to produce new ideas, then (so conventional wisdom has it) they need an economic incentive. Otherwise why would they bother?

Copyright

Copyright was an attempted solution to this problem. It was a legal framework for striking a balance between these two conflicting imperatives: the need to incentivize creative people by enabling them to profit from their creativity; and the need to ensure that society had an abundant supply of ideas. The way it worked was that the creative person was granted a time-limited monopoly on the right to create and sell copies of her work. But after the time limit had expired, the work entered the public domain, and people were free to do whatever they wished with it.

Two things are worth noting about copyright. The first is that it is not a 'right' in the sense that, say, human rights are – that is to say freedoms to which every human being is entitled simply by virtue of being human; copyright is a *monopoly granted by a government*. Secondly, the *monopoly is not a permanent one*: it expires after a defined period of time. As Article 1, Section 8 of the US Constitution puts it, Congress was empowered 'To promote the Progress of Science and useful Arts, by securing for limited Times to Authors and Inventors the exclusive Right to their respective Writings and Discoveries.'

At the time the Constitution was enacted in 1787, the law of copyright regulated only publishers – it applied only to 'maps, charts and books'. This meant that every other aspect of creative life was free.

> Music could be performed in public without a license from a lawyer; a novel could be turned into a play even if the novel was copyrighted. A story could be adapted into a different story; many were, as the very act of creativity was understood to be the act of taking something and re-forming it into something (ever so slightly) new. The Public Domain was vast and rich – the works of Shakespeare had just fallen from the control of publishers in England; they would not have been protected in the United States even if they had not.[5]

The original copyright term envisaged by the men who drafted the US Constitution was fourteen years, with the possibility of one extension of fourteen years, making a total of twenty-eight years in all. But over the years, under pressure from publishers and their lobbyists, Congress steadily increased the term of the monopoly to its present length: life of the author plus seventy years.[6] And although there are differences in detail in different legal jurisdictions, copyright terms across the world have broadly followed the US pattern. In the UK, for example, the current term is also life of the author plus seventy years. What this means is that the freedom of ideas – or at least their expressions in creative works – to mate is severely restricted.

Property and 'intellectual property'

Ideas can be expressed in several forms, of which writing, music and movies are just three. They are also expressed as patents and trademarks – both of which have come to enjoy the protection of time-limited, government-granted monopolies.[7] In what must have looked – at least to lexicographers and lawyers – as a tidying-up operation, these were lumped together with copyright into a grand portmanteau term – *intellectual property*.

On the face of it, this looks like a sensible move, but in fact it has muddied the waters so much that a sensible debate about copyrights – and 'copywrongs', by which I mean the abuse of copyright – is virtually impossible. The difficulty is caused by the word 'property', a term with powerful everyday connotations. Most of the time it means something tangible and real: a building, a parcel of land, a physical possession, so if someone says 'I own this land' it means that (at least in jurisdictions governed by the rule of law) she has a right to the exclusive enjoyment of 'her' property for as long as she lives or continues to own it.

But 'intellectual property' is not property in the same way, for two reasons. Firstly, it is actually a time-limited monopoly granted by the state: it is not a possession that can be held in perpetuity. And secondly, it differs from 'real' property because it is what economists call non-rivalrous. If you purchase or forcibly take possession of my house, then I can no longer use or enjoy it: my house is rivalrous. But if you take my ideas I can still use, enjoy and perhaps profit from them.

One of the most elegant articulations of the non-rivalrous nature of ideas was uttered by Thomas Jefferson in a letter to his friend Isaac McPherson in 1813:

> If nature has made any one thing less susceptible than all others of exclusive property, it is the action of the thinking power called an idea, which an individual may exclusively possess as long as he keeps it to himself; but the moment it is divulged, it forces itself into the possession of every one, and the receiver cannot dispossess himself of it. Its peculiar character, too, is that no one possesses the less, because every other possesses the whole of it. He who receives an idea from me, receives instruction himself without lessening mine; as he who lights his taper at mine, receives light without darkening me. That ideas should freely spread from one to another over the globe, for the moral and mutual instruction of man, and improvement of his condition, seems to have been peculiarly and benevolently designed by nature, when she made them, like fire, expansible over all space, without lessening their density in any point, and like the air in which we breathe, move, and have our physical being, incapable of confinement or exclusive appropriation. Inventions then cannot, in nature, be a subject of property.[8]

Copyright vs. technology

In the scribal age copying wasn't much of a problem, because the act of making a copy was almost as laborious and expensive as that of making the original. But from Gutenberg onwards, copyright has always been, effectively, at war with technology. It's no accident that the question of 'ownership' of ideas expressed in written form first arose as the printing press spread across Europe. Suddenly, it was easy to take someone's work and stamp out thousands of copies without their permission or even their knowledge. Succeeding centuries have seen an intensifying conflict between new technologies that make copying easier, cheaper and more accurate on the one hand, and those who seek to protect their intellectual 'property' as expressed in various media, on the other. In an Internet age, that conflict has become pathological.

When media were analogue, the threat to copyright holders was limited by nature because analogue copying technologies are intrinsically degenerative. If you make a photocopy of this page, then the quality of the copy will be slightly inferior to that of the original. If you make a copy of the photocopy, then the second copy will be, in turn, slightly degraded. By the hundredth copy you'll really start to notice the difference.

The same goes with analogue tape recordings of music, or recordings of movies or TV shows made using video-cassette recorders (VCRs). The quality of the copy was always inferior to that of the original, and successive copying meant increasingly degraded copies. And once you had made an analogue copy, how could you share or distribute it? You could give it to a friend, obviously, or sell it on the black market. But you couldn't distribute it on a global – or even a large – scale. So, in a way, analogue copying was a self-limiting activity

But once digital technology arrived, all that changed. Why? Because once books, recorded music and video are encoded digitally, they simply become streams of binary digits (bits), i.e. ones and zeroes. And copying then simply involves replicating those bitstreams, with the added advantage that there is a simple mathematical test[9] for checking that the copy is a perfect replication of the original. In contrast with analogue technology, where copies can be of different quality, a digital copy is either perfect or it is useless. And the thousandth – or the millionth – digital copy will also be a perfect replication of the original.

The digital computer is, essentially, a copying machine. That's not because it was *designed* to make copies but because it actually works by continually making copies of bitstreams, manipulating them and moving them from one internal register to another. Copying is also an intrinsic part of most interactions with the Internet. When you click on a web link, for example, what happens is that the server on which the requested page resides dispatches a copy of the page – encoded as ones and zeroes – across the Net. When your computer receives the bits, it copies them faithfully into its video RAM, which then enables the machine to display the page on your screen. So the very act of viewing a web page actually requires making a perfect copy of it. Copying is to digital computing, therefore, as breathing is to animal life – in that one cannot exist without the other.

Once the Net – which, after all, is a global machine for transferring bitstreams from one place to another – arrived, we had the makings of a perfect storm for some copyright owners. For not only had each PC user a machine for making perfect copies, but s/he then had access to a global distribution system for 'sharing' those copies. As we saw earlier, the Napster file-sharing system provided the tool needed to make this easy to do. The result was a dramatic upheaval in the music industry and – ultimately – its surrender of effective control to a computer company (Apple) which had figured out a way of legally distributing music tracks.

The story of what digital technology did to the music industry is a salutary case study in how the Net can undermine a cartel of complacent incumbents. A sports reporter would summarize it as: Internet 1, Copyright owners, nil. An MBA student might cite it as an example of what happens when an industry focuses on its own needs (protecting an obsolete business model based around music encoded on plastic disks) rather than attending to the demands of its customers. A lawyer would regard it as a case of widespread infringement of intellectual property. A lobbyist would spin it as 'theft'. But as a guide to the future it isn't really all that interesting: illicitly making copies and distributing them for free isn't, at base, a very creative activity because it doesn't address the problem of how to create new works or art – or at any rate, how to fund the musicians who make the art that is being so freely shared.[10] In that sense, we may regard the file-sharing episode as Round One in the battle between digital technology and copyright.

Round Two is more interesting – and much more important. And it's just begun.

What's driving it is the fact that computers are not just tools for creating perfect copies; they are also tools for creating new works. In the last decade the computing industry has democratized access to powerful software tools for creating, editing and manipulating digital content, and the Internet has provided ordinary people with ways of publishing what people produce with those tools.

These tools include:

- Photoshop, Lightroom, Aperture and the GIMP: programs

which enable users to edit, mix and manipulate images

- Audacity and other tools for audio editing
- iMovie, Adobe Premiere, Final Cut: programs which enable amateurs to edit movies
- GarageBand and other tools for creating and mixing music
- Online services like Wordpress, Blogger, LiveJournal and Typepad which enable anyone who can type to create and publish blogs and other types of website.

Some of these tools are Open Source programs available as free downloads or as web services; some are bundled by computer manufacturers with every new computer or operating system release; and some of those which are sold at commercial prices are relatively affordable. The important thing from our point of view is how powerful and capable they are. On my Apple desktop machine, for example, I have Final Cut Pro[11] – a video editing program that is often used to edit broadcast TV programs and feature films. Twenty years ago, software like this cost thousands of dollars and was available only to professional studios; now versions of these programs are within the reach of most people who can afford a laptop computer.

These tools are the engines that are powering a new explosion of cultural production in which amateurs – in the original sense of the word, i.e. people who do things for the love of it rather than for financial reward – are creating new works either *ab initio*, or by taking existing digital content and reworking it into new forms.

A classic example is a YouTube video[12] satirizing the close relationship between President George W. Bush and Tony Blair, the British prime minister – the politicians who in 2003 launched the invasion of Iraq. The video – which was made by Johan Söderberg, an experienced editor who has edited music videos for Robbie Williams and Madonna[13] – takes TV footage of both leaders and cleverly lip-syncs them to a recording of Lionel Richie and Diana Ross singing Richie's song *Endless Love*.

'*My love*,' croons Bush/Richie, '*There's only you in my life. The only thing that's bright.*'

'*My first love*,' sings Blair/Ross in reply, '*You're every breath that I take. You're every step I make*.'

Well, you get the idea. It's clever and funny and it makes a serious point – that the relationship between two powerful politicians who took their countries to war on the basis of dubious intelligence about weapons of mass destruction can be represented as an outbreak of puppy love.

There's a lot more of this kind of stuff on the Net. Some of it is tiresome, banal and rather boring. But much of it is – as this is – amusing and original. In order to make his clever spoof, however, Söderberg used copyrighted material – video footage from television networks, and a copyrighted recording of a popular song. And in doing so he rendered himself liable for prosecution for copyright infringement.

The Bush-Blair love duet is a paradigmatic example of what Lawrence Lessig calls 'remix culture'.[14] To us, it looks like something new, but really it's just a contemporary manifestation of something that is as old as art itself.

You think I jest? Well, consider one art form that we nowadays revere – classical music. Many of the people we regard as the greatest composers in the classical canon were inveterate 'borrowers' and remixers. J.S. Bach, for example, borrowed from Telemann, Frescobaldi, Albinoni and Vivaldi. Haydn used melodic material from various sources. And even Mozart borrowed from, and was influenced by, other composers, including Haydn and Bach.[15] Handel, for his part, is celebrated by musicologists for his extensive borrowing and reworking of other people's works. 'He would,' writes Paul Henry Lang, 'take a small segment, often no more than an incipit that none of us would dignify with a second look, and make it the germinal element of a tremendous musical edifice of his own. One marvels at the small alterations that would make a borrowed element coming from a totally different environment exactly right for the occasion.'[16]

And in the twentieth century, the American composer Charles Ives was an inveterate user of borrowed material. According to one analysis, almost two hundred individual works or movements by Ives, spanning his entire career and representing more than a third of his output, incorporate music by other composers or from his own previous work.[17]

This theme – of borrowing and remixing as an intrinsic part of cultural creativity – can be found in the history of most other major art forms: painting, literature, dance, to name but three. The great irony, though, is that if our current copyright laws had existed when Bach, Mozart, Haydn and Handel were in their primes, none of it would have been allowed. Instead of composing at the furious rate that they did, all four would have been up to their eyes in cease-and-desist letters from expensive lawyers in wigs and frock coats. And if that had indeed happened, would posterity – that's you and me – be richer or poorer as a result?

This is not an attack on copyright, by the way, but an observation on the way in which the original balance between the rights of content-owners and the needs of society has become seriously skewed in the post-war era. In his book on remix culture,[18] Larry Lessig tells a story that illustrates the extent of the skew.

It concerns the experience of Stephanie Lenz, an American mother who posted to YouTube a touching little video clip of her eighteen-month-old son moving rhythmically to the beat of a song by the singer Prince. A few months later she was astonished to receive a notice from YouTube informing her that the company had received a 'takedown' notice from Universal Music Group on the grounds that the clip had infringed Prince's copyright. Ms Lenz took the case to the Electronic Frontier Foundation, a not-for-profit body that campaigns for freedom in cyberspace. The Foundation's lawyers filed a routine counter-notice, arguing that no rights of Prince had been violated and that Ms Lenz had the right to show her baby's dancing skills, and assumed that that would be the end of the matter. But it wasn't: Universal fought back. Since then the case has been in and out of the courts, and at the time of writing does not seem to have been resolved.[19]

But the video was still available on YouTube when I last checked (and it had been seen by more than a million people).[20] It captures a lovely moment in a family's life and evokes the blissful innocence of young children. But what's really striking is how poor the quality of the audio is. Most people coming to the clip without knowing the background would be unable to identify the track to which the toddler was responding. The idea that this particular re-use of a song might somehow cause the singer to lose revenue is, frankly, absurd.

What the case of Lenz vs. Universal illustrates is how distanced copyright law has become from the realities of networked life. 'Imagine,' writes Lessig,

> the conference room at Universal where the decision was made to threaten Stephanie Lenz with a federal lawsuit. Picture the meeting: four, maybe more, participants. Most of them lawyers, billing hundreds of dollars an hour. All of them wearing thousand-dollar suits, sitting around looking serious, drinking coffee brewed by an assistant reading a memo drafted by a first-year associate about the various rights that had been violated by the pirate, Stephanie Lenz. After thirty minutes, maybe an hour, the executives come to their solemn decision. A meeting that cost Universal $10,000? $50,000? (when you count the value of the lawyers' time, and the time to prepare the legal materials); a meeting resolved to invoke the laws of Congress against a mother merely giddy with love for her eighteen-month-old.[21]

So it is that we find ourselves in a very strange and contorted position. *Copyright – which, after all, was created to encourage and reward creativity – is now being routinely used to stifle it.* How did it come to this?

Read-only culture and its consequences

Once upon a time, all music involved being present at a performance, either as a listener or a music-maker. But Thomas Edison changed all that. As Evan Eisenberg put it in his classic study,[22] when the phonograph arrived music became 'an object that could be owned by the individual and used at his own convenience'. The significance of this shift was missed by most people, caught up as they were in the excitement of this wonderful new 'recording' technology. But some people understood what was happening. The composer John Philip Sousa was one of them. While testifying before Congress about his concerns about copyright, he said something that for many years seemed quaint and now seems suddenly relevant. 'When I was a boy,' he mused,

> in front of every house in the summer evenings you would find young people together singing the songs of the day or the old songs. Today you hear these infernal machines going night and day. We will not have a vocal chord left. The vocal chords will be eliminated by a process of evolution as was the tail of the man when he came from the ape.[23]

Well, we still have our vocal chords, but Sousa was onto something, because the history of the twentieth century shows how entertainment changed from being something one *did* to something that one *consumed*. His fear, writes Lessig,

> was not that culture, or the actual quality of the music produced in a culture, would be less. His fear was that people would be less connected to, and hence practiced in, creating that culture. Amateurism, to this professional, was a virtue – not because it produced great music, but because it produced a musical culture: a love for, and an appreciation of, the music he re-created, a respect for the music he played, and hence a connection to a democratic culture. If you want to respect Yo-Yo Ma, try playing a cello. If you want to understand how great great music is, try performing it with a collection of amateurs.[24]

In computerspeak the culture that Sousa was extolling would be called a read-write one. The metaphor derives from the way in which permissions are assigned to files in a computer operating system. If the user has R/W permission for a particular file s/he is allowed to both read it and make changes to it. This contrasts with R/O (read-only) permissions, which only allow the user to read the contents of the file.

Before the invention of the phonograph, musical culture in most parts of the world was a read-write one. As Henry Jenkins puts it, 'the story of American arts in the nineteenth century might be told in terms of the mixing, matching and merging of folk traditions taken from various indigenous and immigrant populations.'[25] But as the new communications technologies of the twentieth century emerged, blossomed and took hold, popular culture steadily degenerated into a read-only state. It started with pianolas (so-called player-pianos driven by paper rolls in which music was encoded by patterns of punched holes), gathered force with the phonograph (gramophone), movies, and the broadcast (i.e. few-to-many) media of radio and television. As Lessig argues, the twentieth century was the first time in the history of human culture 'when popular culture had become professionalized, and when the people were taught to defer to the professional'[26]

The significance of this is that it was in this read-only culture that our current thicket of copyright laws evolved. As Tim Wu and Yochai Benkler have pointed out, achieving the dominance needed to be successful in a read-only, mass-media-dominated culture required the deployment of vast amounts of capital – in recording and movie studios, broadcasting equipment and licences, complex distributions systems, advertising networks and so on. Investment on this scale meant that powerful industrial lobbies emerged to protect the interests of media owners and corporations, and the inexorable distortion of the original balance in copyright law between the rights of content-creators and the needs of society was a predictable outcome. So the intellectual property laws we currently have in relation to content are bespoke products which were exquisitely crafted to protect the interests of those who were the dominant forces in our old media ecosystem.

The real war: the future vs. the past

But – as we know – that ecosystem is changing. Computing and the Internet have granted read-write permissions to ordinary people – to amateurs. The explosion in user-generated creativity – as expressed in blogs, Flickr, mashups, remixes, YouTube spoofs, websites, free music, Wikipedia and so on – is the result. This is read-write culture, and underpinning it is a technology that makes perfect copying not just possible but trivially simple. Yet we are trying to impose on it laws that define the making of any unauthorized copy as a crime.

This is bad news for two reasons. Firstly, it offends against common sense to try and implement laws designed for a different age. In that sense, it calls to mind a celebrated legal case that was decided by the US Supreme Court just after the Second World War. At its heart was the question of what rights ownership of land confers on landowners. For centuries, conventional legal opinion held that a property owner owned not just the surface of his land, but all the land below, down to the earth's core, and all the space above to 'an indefinite extent, upwards'. This doctrine was not just theoretical; it provided the basis, for example, on which British landowners were able to exploit the coal which lay beneath their estates. So it was a legal doctrine with important economic consequences.

But what about the landowner's ownership of the space *above* his property? The issue came to a head with the invention of the aeroplane. Two brothers, Thomas Lee and Tinie Causby, who were chicken farmers in North Carolina, began to lose some of their chickens when low-flying military aircraft passed over their farm. It seems that the unfortunate fowl panicked, flew into the walls of the barn, and killed themselves in the process. So the brothers Causby filed a legal suit against the government, claiming that it was trespassing on their property.

The case reached the Supreme Court on 1 May 1946 because it had constitutional implications. Congress had declared the airways to be public property, but if property rights really extended to the heavens, then Congress's declaration could have represented unconstitutional appropriation of property without compensation. The court acknowledged the ancient doctrine that ownership of land extended to 'the periphery of the Universe' but decided – by a majority of five to two – that such a doctrine could not be allowed to rule indefinitely. The doctrine, said Justice William O. Douglas (who wrote the ruling opinion),

> has no place in the modern world. The air is a public highway, as Congress has declared. Were that not true, every transcontinental flight would subject the operator to countless trespass suits. Common sense revolts at the idea. To recognize such private claims to the airspace would clog these highways, seriously interfere with their control and development in the public interest, and transfer into private ownership that to which only the public has a just claim.[27]

The phrase that leaps from the page is Douglas's terse summary: *common sense revolts at the idea*. It does. And it should also revolt at the idea that doctrines about copyright that were shaped in a pre-Internet age should apply to a post-Internet one.

But there is another, and perhaps more compelling, reason, for rethinking our approach to copyright in this new context. It is that the laws in their present form are effectively unenforceable at any significant scale. Representatives of the content industries will not concede this, of course, for understandable reasons: nobody wants to admit that they are waging an unwinnable war. They bluster on about Digital Rights Management (DRM) technologies for locking down content and restricting what users can do with it, but I don't know any computer scientist who believes that an unbreakable DRM system is a realistic possibility.

The content industries know that too, of course, which is why they thought it necessary to have a second legal weapon up their collective sleeve – the US Digital Millennium Copyright Act (DMCA) of 1998 which, among other things, makes it a criminal offence to circumvent anti-copying measures implemented by copyright holders.[28] So – the industry thinking goes – we can lock down our content using DRM, and if people break the DRM the DMCA will ensure that they wind up in jail. QED.

Ponder for a moment where this line of thinking is headed. Although there are lots of isolated examples of the DMCA being deployed successfully against individuals and websites, there is no evidence that the unauthorized copying of digital content is in decline. *Au contraire*: just look at the amount of stuff that is being posted to YouTube and at what goes on all over the Web. Online downloading, borrowing and remixing are now a normal part of youth culture. It's become a way of life.

So we're facing a situation where large numbers of our fellow citizens are effectively being criminalized by unenforceable laws. Does that remind you of anything? As luck would have it, we know what happens when an unenforceable law tries to regulate behaviour that millions of people regard as pleasurable and normal, because the US tried that experiment between 1920 and 1933. It was called Prohibition. During that thirteen-year period the sale, manufacture, and transportation of alcohol were banned in the whole of the United States. While the legislation did succeed in reducing the amount of alcohol consumed, it led to widespread flouting of the law by ordinary citizens and a massive expansion in organized crime. In the end, the experiment was terminated with the ratification of a constitutional amendment repealing the earlier amendment which had given rise to the banning of alcohol in the first place.

Are we going to try the same approach with copyright? And if we do, is there any reason to suppose that the results will be any different?

9. ORWELL VS. HUXLEY: THE BOOKENDS OF OUR NETWORKED FUTURE?

There are, wrote the cultural critic Neil Postman, 'two ways in which the spirit of a culture may be shrivelled. In the first – the Orwellian – culture becomes a prison. In the second – the Huxleyean – culture becomes a burlesque'.[1] Although the focus of Postman's attention was US broadcast television and – like most writers at the time (the mid-1980s) – he was probably oblivious to the Internet, it's always seemed to me that his juxtaposition of these two celebrated novelists provides an insightful way of thinking about our networked future. Orwell thought that we would be destroyed by the things we fear, while Huxley's nightmare was that we would be undone by the things we enjoy. The strange thing is that, in an Internet-centric world, these two – ostensibly incompatible – forces might combine to transport us to states that both writers would recognize, and that we might come to regret.

But first, a caveat. For me, Postman is the most compelling media commentator of my lifetime, but it's important to remember that he operated and wrote in a particular historical context. He was at the height of his powers just when the old, TV-dominated ecosystem had reached its peak, and he was a baleful, unrelentingly hostile, critic of the medium, which he saw as a force that was not just dumbing down American society, but infantilizing it. He was also a formidable critic of what he called 'technopoly', that is to say the conflation of technology and ideology.[2] But – surprisingly, given that he died in 2003 and was still writing and publishing in the 1990s – Postman wrote almost nothing about the Internet, and indeed seems to have engaged only peripherally with it.[3] So in borrowing his metaphor about Orwell and Huxley, I'm taking it beyond the context in which he embedded it.

The Orwellian nightmare

George Orwell (1903–50) was a journalist, critic and novelist of immense range and power, who is still revered by many people as a sea-green incorruptible – a writer who told the truth as he saw it, no matter how inconvenient it might have been to those in power.[4] He is particularly remembered for two dystopian novels, *Animal Farm* and *Nineteen Eighty-Four*, in which he outlines nightmarish visions of how our societies might evolve. *Animal Farm* is a parable about animals who revolt against the oppression of their human masters, but discover that, even in a world where all animals are supposed to be equal, some (in this case the pigs) are more equal than others.

In *Nineteen Eighty-Four*, which was written in 1947–8 and published in 1949, Orwell sketches out his vision of what a totalitarian society would be like. It tells the story of Winston Smith, a civil servant in the fictional society of Oceania in which the individual is always subordinated to the state (embodied by 'The Party') which is characterized by perpetual war, pervasive government surveillance by 'Big Brother', and incessant public mind control. As a civil servant in the Ministry of Truth, Smith is responsible for perpetuating the Party's propaganda by revising historical records to render it apparently omniscient. Smith rebels against this regime, which leads to his arrest, torture – and finally to his re-conversion to the Party's view of the world.

The society envisioned in *Nineteen Eighty-Four* seems to have been inspired by two factors: the rise of Nazi Germany and the evolution of the Soviet Union under Stalin. Orwell may also have been influenced by his experiences of living in wartime Britain, a democratic society in which – under the pressures of war – the state had appropriated to itself more or less untrammelled power. Since the Nazis were defeated, and the Soviet Union imploded, the novel may seem rather dated to modern readers, and yet there are aspects of it which still resonate with us.

One measure of Orwell's significance as a writer is the way some of his ideas have found their way into everyday language. The phrase 'Big Brother', for example, is universally used to conjure up images of comprehensive and intrusive surveillance. So are other terms from *Nineteen Eighty-Four* – 'Newspeak', 'Doublethink' and 'Thought Police'. More generally, the adjective 'Orwellian' evokes 'an attitude and a policy of control by propaganda, surveillance, misinformation, denial of truth, and manipulation of the past, including the "unperson" – a person whose past existence is expunged from the public record and memory, practised by modern repressive governments.'[5]

The aspect of Orwell's vision that has achieved a startling new relevance for a networked society is his idea of total surveillance. On a wall in Winston Smith's apartment is 'an oblong metal plaque like a dulled mirror which formed part of the surface of the right-hand wall'. This is the telescreen, an instrument which is found in every apartment in Oceania and which could be dimmed but never switched off completely.

> The telescreen received and transmitted simultaneously. Any sound that Winston made, above the level of a very low whisper would be picked up by it; moreover, so long as he remained within the field of vision which the metal plaque commanded, he could be seen as well as heard. There was of course no way of knowing whether you were being watched at any given moment. How often, or on what system, the Thought Police plugged in on any individual wire was guesswork. It was even conceivable that they watched everybody all the time. But at any rate they could plug in your wire whenever they wanted to. You had to live – did live, from habit that became instinct – in the assumption that every sound you made was overheard, and, except in darkness, every movement scrutinized.[6]

Quaint, isn't it? Well try this: point your web browser at www.whatismyip.com and see what happens. The site returns a set of four numbers, separated by full stops. That's the Internet Protocol (IP) address of your computer or (if you're on a network) of the router which connects your network to the Internet. In everyday terms, it's the unique address of the computer or router.[7]

Now comes the Orwellian bit. Everything that uniquely addressed computer does – every email it dispatches or receives, every web page it requests and views, every Instant Message or Facebook update or Tweet is logged and recorded at your ISP and at various places across the Internet. The moment you click on a web link, for example, your computer will reveal to the target website:

- The IP address of your machine (or network router). From this the receiving machine can infer your physical location, or at least the physical location of your ISP.
- What kind of computer you are using, including details of the operating system and the web browser.
- If you have visited the website before, it may have deposited a 'cookie' – a small text file – on your hard disk which it will then read to extract information about your previous visits, preferences, etc.

So everything you do in cyberspace leaves a trail, including the '**clickstream**' that represents the list of websites you have visited, and anyone who has access to that trail will get to know an awful lot about you. They'll have a pretty good idea, for example, of who your friends are, what your interests are (including your political views if you express them through online activity), what you like doing online, what you download, read, buy and sell. You can, if you are technically adept, cover some of your digital tracks by using anonymizing techniques,[8] but for most Internet users life is too short for such measures, and so they cheerfully roam about cyberspace, leaving a rich spoor of digital footprints wherever they go.

For governments of all political stripes – from authoritarian regimes to liberal democracies – the Internet is a surveillance tool made in heaven, because much of the surveillance can be done, not by expensive and fallible human beings, but by computers. It can be automated, in other words. The network's usefulness for these purposes was limited in the early days, because it was just a minority sport for the élites of the industrialized world. But its utility has increased dramatically as the number of Internet users – currently amounting to nearly a third of the world's population – has increased.

'While the internet has in some ways an ability to let us know to an unprecedented level what government is doing and to let us co-operate with each other to hold repressive governments and repressive corporations to account,' said Julian Assange, the founder of WikiLeaks, in a speech to Cambridge students in March 2011,

> it is also the greatest spying machine the world has ever seen. It is not a technology that favours freedom of speech. It is not a technology that favours human rights. It is not a technology that favours civil life. Rather it is a technology that can be used to set up a totalitarian spying regime, the likes of which we have never seen.[9]

Governments employ different rationales for Internet surveillance. Authoritarian regimes see it as an essential way to detect and eliminate political dissent. Democratic administrations maintain that Internet monitoring is essential to combat terrorism and organized crime. (And claim that judicial and other safeguards are in place to limit the possibility of abuse.) But whatever the rationale, the fact is that nowadays they are all in the surveillance business. In the United States for example, under the Communications Assistance for Law Enforcement Act,[10] all telephone calls and broadband Internet traffic (emails, **clickstreams**, instant messaging, etc.) are required to be available for unimpeded real-time monitoring by federal law-enforcement agencies. Under the same act all US telecommunications providers are required to install so-called 'packet sniffing' technology[11] to allow federal law enforcement and intelligence agencies to intercept all of their customers' broadband Internet traffic.[12] In the UK, the Regulation of Investigatory Powers Act[13] confers analogous powers on the British authorities, and the country's intelligence services have unveiled plans to store records of every online communication.[14] Other European countries have similar kinds of enabling legislation and/or plans.

So in a strange way, Orwell's telescreen has morphed into the LCD screens of our desktops, laptops, tablets and smartphones. Big Brother may not be watching us all the time, but the Internet has given him the tools he needs for the job, and we have no idea how assiduous he is in monitoring us.

Huxley's world

There's a nice irony in the fact that the book-ends of our future identified by Postman should both be English writers who were alumni of the same educational establishment – Eton College, Britain's most exclusive private school. Not only that, but Aldous Huxley (1894–1963) was a teacher at Eton during Orwell's first year (1917–18) as a pupil there.[15]

Huxley was born into England's intellectual aristocracy. His grandfather was Thomas Henry Huxley, the Victorian biologist who was the most effective public evangelist for Darwin's theory of evolution.[16] His mother was the niece of the poet and essayist Matthew Arnold, and he was the nephew of the celebrated Victorian novelist Mrs Humphrey Ward. His brother Julian and half-brother Andrew both became distinguished biologists. In the circumstances, it's not surprising that Aldous turned out to be a writer who ranged far beyond the usual preoccupations of literary folk – into history, philosophy, science, politics, mysticism and psychic exploration. His biographer wrote that 'he offered as his personal motto the legend hung around the neck of a ragged scarecrow of a man in a painting by Goya: *aun aprendo*. I am still learning.'[17] He was, in that sense, a modern Voltaire, a genuine *philosophe*.

Brave New World, his most celebrated novel, was written in 1931 and published in 1932. The title comes from a phrase in one of Shakespeare's plays.[18] It is set in the London of the distant future – 2540 AD – and describes a society based on Huxley's imaginative extrapolation of scientific and social trends. In a letter to his father, he described the book as

> a comic, or at least satirical, novel about the Future, showing the appallingness (at least by our standards) of Utopia and adumbrating the effects on thought and feeling of such quite possible biological inventions as the production of children in bottles (with consequent abolition of the family and all the Freudian 'complexes' for which family relationships are responsible), the prolongation of youth, the devising of some harmless but effective substitute for alcohol, cocaine, opium, etc. and also the effects of such sociological reforms as Pavlovian conditioning of all children from birth and before birth, universal peace, security and stability.[19]

Brave New World is shaped by several of its author's obsessions. When he first visited the US Huxley was stunned by the way in which the distractions offered by an affluent consumer society could dull the critical faculties of citizens. He was particularly struck by the way in which a population could apparently be rendered docile by advertising and retail therapy. As an intellectual who was fascinated by science, he guessed (correctly, as it turned out) that scientific advances would eventually give humans powers that had hitherto been regarded as the exclusive preserve of the gods. And finally, his encounters with industrialists like Alfred Mond led him to the idea that societies would eventually be run on lines inspired by the managerial rationalism of mass production ('Fordism') – which is why the year 2540 AD in the novel is 'the Year of Our Ford 632'.

These obsessions converged in *Brave New World*. In it, Huxley describes what amounts to the mass-production of children by an industrialized variation on what we would now call *in-vitro fertilization*; interference in the development process of infants to produce a number of 'castes' with carefully modulated levels of capacity to enable them to fit without complaining into the various societal and industrial roles assigned to them; and Pavlovian conditioning of children from birth. It's a world in which nobody falls ill, everyone has the same life-span, there is no warfare, and institutions and mores like marriage and sexual fidelity are dispensed with. Huxley's is a totalitarian society, ruled by a supposedly benevolent dictatorship whose subjects have been programmed to enjoy their subjugation through conditioning and the use of a narcotic drug – Soma – which is less damaging and more pleasurable than any narcotic known to us. The rulers of *Brave New World* have solved the problem of making people love their servitude.

Like *Nineteen Eighty Four*, *Brave New World* is a parable about what could happen to us and to our societies. In both cases, the parables are about how human beings might be controlled. Orwell's version was informed by knowledge of what Nazi and Soviet totalitarianism had achieved and by his intuitions about what it would be like to live in a society dominated by fear. Huxley's version was informed by the insight that conditioning by positive reinforcement was likely to provide a more successful way of controlling people. 'In *Nineteen Eighty Four*,' he wrote later, 'the lust for power is satisfied by inflicting pain; in *Brave New World*, by inflicting a hardly less humiliating pleasure.'[20]

It's easy to see why Neil Postman felt that Huxley's dystopia represented the greater long-term threat to freedom. As the title of Postman's best-known book (*Amusing Ourselves to Death*) implied, he was convinced that American television represented a mortal threat to society. 'In the Huxleyan prophecy,' he wrote,

> Big Brother does not watch us, by his choice. We watch him. There is no need for wardens or gates or Ministries of Truth. When a population becomes distracted by trivia, when cultural life is defined as a perpetual round of entertainments, when serious public conversation becomes a form of baby-talk, when, in short, a people become an audience and their public business a vaudeville act, then a nation finds itself at risk: culture-death is a clear possibility.[21]

What has all this to do with the Net?

At first sight, very little. Postman was writing about an information ecosystem dominated by broadcast media, whereas it looks as though the power and influence of those media will be significantly reduced in an Internet-centric ecosystem. Surely the remarkable proliferation of blogging and other kinds of user-generated content – the explosion in what Clay Shirky calls 'the mass amateurization of publishing'[22] – has widened the public sphere, challenged the dominance of existing media organizations, increased the diversity of public expression and generally undermined the power of the broadcasters whom Postman fingered as the causes of cultural decline? Surely the nightmare vision of a dumbed-down society of couch potatoes has receded as the Internet has gone mainstream?

The answer is that it undoubtedly has – *up to now*. The Internet *has* liberated a great deal of previously untapped human creativity, and provided its users with a cornucopia of delights – an unimaginably wide range of opportunities for entertainment, information and communication. But concealed in those delights there may be a bitter pill that Huxley would have recognized, for in pursuing the things we enjoy we could, in fact, be sleepwalking into a kind of pleasurable servitude.

How come? Well, here are two thoughts that give one pause.

The tyranny of power laws

The first is the realization that the cumulative outcome of millions of independent decisions might not, paradoxically, be correspondingly greater diversity of choice. This was one of the most startling discoveries of the early studies of the Web. In most areas of life there's a reassuring regularity about the frequency with which various properties occur. Take people's height, for example. If you were to measure the height of a thousand people, then you'd be able to work out the average (or mean) height, and you'd find that most people are of average height, and that very small or very tall people are rare, with a sliding scale in between.

The intriguing thing about the Web is that this kind of

distribution doesn't seem to apply to it. Instead, the distribution of visits and links to websites follows what statisticians call a *power law*.[23] Instead of the vast majority of sites having an average number of links and visits, what you find is that a small number have huge numbers, while an enormous number have very few. So the statistical distribution of visits or links has a large peak on the left, and a long 'tail' stretching out to infinity on the right.

Ponder, for a moment what this means. There are millions of websites out there, and nearly two billion Web users. Each user is sovereign, in the sense that s/he is free to click on any link, or to create a link to any other page on the Web. And yet, in the end, the aggregate result of all those billons of freely made, independent choices is that a relatively small number of websites get most of the links and attract the overwhelming volume of traffic, while the vast majority of sites attract only a relatively small number of visits and links. Which implies two things: firstly that while the range of options available to Web users may be infinite in theory, in *practice* it turns out to be much more constrained; and secondly that if you own or control one of the big sites, then you wield a corresponding amount of power and influence.

In January 2011, the top ten websites in the world in terms of number of visits were[24]:

- Google (search engine and other services, including webmail: main site)
- Facebook (social-networking site)
- YouTube (video hosting site; owned by Google)
- Yahoo (web portal, search engine, webmail and other services)

- Windows Live (online software and web services; owned by Microsoft)
- Baidu (Chinese search engine)
- Blogger (blog-hosting platform; owned by Google)
- Wikipedia (user-generated and moderated online encyclopedia)
- QQ (Asian Internet portal with heavy penetration into the Chinese market)
- Twitter (micro-blogging site)

In the same month, of the top twenty sites in the world, six were owned by Google, three by Microsoft and two by Yahoo. And also in that month, 73.5 per cent of the world's Internet users visited either Google's main site or YouTube, its video-hosting subsidiary. Of the top twenty sites in the world, all but three were owned by US-based companies.[25] This represents an extraordinary concentration of attention on the part of a huge global audience. We use the services of these companies with gleeful abandon and forget that, in the process, they are learning a great deal about us. One day we may come to realize the truth of the old adage: in an information age, *knowledge is power*.

The Wu Cycle

To baby-boomers like me, who were students when the ARPAnet – the Internet's precursor – was taking shape, the network has always had a romantic aura. We believed that engineers had created a communications system that was revolutionary and transformational – something that would change the world for the better. The man who first gave voice to this techno-Utopianism was John Perry Barlow, the lyricist of the Grateful Dead, the celebrated 1960s San Francisco rock band. His celebrated Declaration of the Independence of Cyberspace (quoted in Chapter 1) was a riposte to a particular political initiative aimed at censoring the Internet.

This was the Communications Decency Act (CDA) which the US Congress passed in 1996 and which imposed criminal sanctions on anyone who

> knowingly (A) uses an interactive computer service to send to a specific person or persons under 18 years of age, or (B) uses any interactive computer service to display in a manner available to a person under 18 years of age, any comment, request, suggestion, proposal, image, or other communication that, in context, depicts or describes, in terms patently offensive as measured by contemporary community standards, sexual or excretory activities or organs.[26]

The act also criminalized the transmission of materials that were 'obscene or indecent' to persons known to be under 18.

Parts of the CDA were struck down by the US Supreme Court in 1997, but the very fact that it had even briefly been part of the criminal code had by that stage enraged libertarians everywhere. Barlow's 'Declaration' was his instinctive response to the act. Its stentorian tones and high-flown oratory may seem quaint now, but it expresses very well the spirit of the age. In particular it articulates a conviction that many of us shared at the time: that the Net was truly special – different from anything that had gone before.

Now, a decade and a half later, after re-reading Orwell and Huxley, we need to ask ourselves a tough question: is the Internet really unique in being immune to censorship or control?

One expert who has doubts on this score is Tim Wu, a legal

scholar who has written extensively on intellectual property, telecommunications policy, Internet governance and the doctrine of 'Net neutrality'. In a thoughtful and disturbing book,[27] he has looked at the evolution of other communications technologies in order to see what lessons history might hold for the Internet. What history shows, Wu argues, is that there is a pattern in the development of these technologies. His research reveals a progression of communications technologies from a freely-accessible channel to one controlled by a single corporation or cartel. It is, he writes,

> a progression so common as to seem inevitable, though it would hardly have seemed so at the dawn of any of the past century's transformative technologies, whether telephony, radio, television, or film. History also shows that whatever has been closed too long is ripe for ingenuity's assault: in time a closed industry can be opened anew, giving way to all sorts of technical possibilities and expressive uses for the medium before the effort to close the system likewise begins again.[28]

This pattern, this oscillation of information industries between open and closed, is so pervasive that Wu assigns it a name – the Cycle. For him – as for us – the key question is: does the Cycle apply to the Net?

This is not an 'academic' question, in the sense of one with few practical implications. The Internet is now central to the existence and functioning of all developed countries: we really do live in an 'information society'. In the past we were much less reliant on information than we are today, and we got information via a variety of different channels. Our future, though, is one in which we are much more reliant on information, most of which will travel on a single network. But, muses Wu,

> is almost certain to be an intensification of our present reality: greater and greater information dependence in every matter of life and work, and all that needed information increasingly traveling a single network we call the Internet. If the Internet, whose present openness has become a way of life, should prove as much subject to the Cycle as every other information network before it, the practical consequences will be staggering. And already there are signs that the good old days of a completely open network are ending.[29]

Wu's history of the information industries in the US makes sobering reading. What's particularly striking is the discovery that the early years of most of the communications technologies researched by Wu were accompanied by optimistic hopes or Utopian dreams. Every new communications medium brought with it hopes that it would ameliorate the ills of society. Broadcast radio, for example, attracted 'an extraordinary faith in its potential as the benefactor, perhaps even a savior, of mankind'. The urge to exploit the new medium stemmed from humanitarian as well as economic motives. 'Well before the Internet,' Wu writes,

> in a world without paid downloads, even before commercial television, the same urge to tinker and to connect with others for the pure good of it gave birth to what we now call broadcasting and practically defined the medium in its early years. In the magazines of the 1910s you can feel the excitement of reaching strangers by radio, the connection with thousands and

> the sheer wonder at the technology. What you don't hear is any expectation of cashing in.[30]

In the US, where broadcasting began, people dreamed that it would reduce the distance between citizens and a remote federal government, that it would elevate the level of public and political discourse, and that it would lead to a cultured society.[31] Viewed against this background, the hopes and dreams of the early Internet evangelists seem almost tame.

In the long view of history, therefore, a pattern can be seen. New inventions lead to a period of openness, excitement and a feeling that nothing will ever be the same again. But the openness doesn't last. Closure is triggered by the arrival of one or more charismatic entrepreneurs at the point when the novelty of the new technology is beginning to wane and consumers have developed a taste for quality, stability and higher production values than are being delivered by the nascent industry. The newcomers offer a better proposition: in telephony, for example, AT&T offered a single network (as opposed to the variety of incompatible phone systems then in existence) together with the promise that customers would get a dial tone when they picked up their handsets; in radio, NBC offered better programming, with professional actors, better scriptwriting, and so on; in movies, the emerging moguls, faced with the creative chaos of the silent movie business, built vertically integrated businesses which owned studios as well as cinemas, employed stars, and delivered sound (and, later, colour) – in other words a more attractive, uniform product. And consumers respond to these propositions, which leads to a positive feedback loop: the new entrepreneurs become more and more successful, their competitors fall away and eventually the industry is effectively captured either by a monopolist (telephony), or a cartel (Hollywood).

The insidious thing is that this process of capture (or closure) doesn't involve any kind of authoritarian takeover. It comes, Wu says, not as a bitter pill but as a 'sweet pill, as a tabloid, easy to swallow.' Most of the monopolists of history delivered a product that was better than what went before – which is what consumers go for and what leads industries towards closure.[32]

The new moguls

It is this last insight – that it is consumers' enjoyment of products and services that leads to their enslavement – that brings Huxley's *Brave New World* to mind. Tim Wu's book contains a fascinating parade of the charismatic entrepreneurs who achieved that level of industrial dominance. First up is Theodore Vail, the man who transformed the Bell Telephone Company from an industrial minnow into AT&T, one of the most powerful corporations in history. Other dominant entrepreneurs in his list are David Sarnoff, who created NBC and ran the Radio Corporation of America (RCA) and Adolph Zukor, who founded Paramount Pictures and revolutionized the film industry by organizing production, distribution, and exhibition within a single company.

These kind of entrepreneurs – people who transform entire industries – are a very rare breed. Joseph Schumpeter, the great theorist of capitalist development, regarded them as the essential motors of innovation. Their motivation was, he maintained, not money but 'the dream and will to found a private kingdom'. They have 'the will to conquer: the impulse to fight, to prove oneself superior to others'. And they are turned on by 'the joy of creating' new products and services.

In our time, we have seen a new generation of this rare breed. One thinks of Bill Gates, the founder of Microsoft; of Larry Ellison, the founder of Oracle, the database company; of Larry Page and Sergey Brin, the co-founders of Google; of Mark Zuckerberg, the founder of Facebook. But most of all one thinks of Steve Jobs, the co-founder of Apple and the architect of its resurgence at the end of the twentieth century and the opening decades of the next.

Jobs transformed Apple from a distinctive but failing computer manufacturer into the second most valuable corporation in the world, first by becoming a byword for product design, and then by establishing a choke-hold on the distribution of digital music, and finally by dominating the market for mobile computing devices.

In collaboration with a brilliant engineer, Steve Wozniak, Steve Jobs co-founded Apple Computer in 1976, and in 1978 they launched the Apple II, the world's first consumer-oriented personal computer. In the early 1980s, having spotted the potential of the graphical user interface developed at Xerox's Palo Alto Research Center (PARC), Jobs inspired and drove the development of the Apple Macintosh – the first consumer product to embody such a user interface – which was launched in 1984. Shortly afterwards, after losing a power-struggle on the company's board, Jobs left Apple and founded NeXT, a company devoted to making high-end engineering workstations, and Pixar, a film animation company.

In 1996, a floundering Apple bought NeXT and as a result Jobs returned to the company he had founded. From 1997 until his death in 2011 he has been its CEO, and the person who was identified in the public mind as the personification of the company and its values.

What are these values? Essentially a fanatical obsession with elegance, functionality and design. These things preoccupied Jobs from the beginning: when the Macintosh was being created, in an era when computers came in grey steel boxes, he famously sought inspiration in the kitchenware sections of department stores because of his belief that *personal* computers ought to look like other consumer products, not like machines. 'Back when they were coming up with the design of the computer,' wrote Farhad Manjoo,

> Jobs kept pestering engineers with ever-more-outlandish ideas. First he wanted the Mac to look like a Porsche. Another time, he went to Macy's and spotted a beautiful Cuisinart, and that became the new template for the Mac – now it had to look like a food processor, he told his team. That idea didn't pan out, but everyone knew what Jobs was going for. 'It's got to be different, different from everything else,' Jobs kept saying.[33]

This conviction that computing devices ought to be beautiful objects was accompanied by a belief that they should also be easy and intuitive to use. This was the spirit that underpinned Apple's move into the digital music business. Most people associate this with the launch of the iPod, the first really elegant portable MP3 player, on 23 October 2001. But in fact an equally significant development was the launch, nine months earlier, of iTunes – Apple's software for 'ripping' (i.e. copying and compressing) audio tracks from CDs and storing them on a computer's hard drive. This meant that when the iPod appeared on the market, there already existed a simple, foolproof way of transferring the owner's music collection from hard disk to player. All one had to do was to connect the iPod to one's laptop or desktop Macintosh and the two devices then automatically 'synced' (i.e. synchronized) with one another.

Why is this significant? Well, it comes back to one of the lessons of the history of communications technology as told by Tim Wu. His analysis suggests that pivotal moments in the evolution of an industry come when an entrepreneur arrives to offer consumers higher quality production values and/or greater ease of use than are being delivered by the incumbents. Prior to the arrival of the iPod there were lots of portable MP3 players, and a raft of different programs for ripping, organizing and playing music tracks on computers. But getting tracks from PC to player – and keeping the two devices synchronized – required a certain amount of technical knowledge. Geeks and early adopters love such challenges. But the average consumer does not – which is why, when Steve Jobs launched the iPod, he was pushing at an open door.

Jobs's next step along the path charted by Wu involved bringing legalized order to the chaotic, copyright-infringing bazaar of online music distribution. Faced with the challenge of illicit file-sharing, the music industry had made a few cack-handed attempts to provide a system for purchasing music tracks, but none had worked. Jobs, sensing that there was a public appetite for high-quality, legally purchased music, modified iTunes to include a seamless gateway to an online store, and persuaded a number of record labels to sell music tracks and albums through it. Apple provided the networking infrastructure to make this work, and took a commission on every track sold.

The iTunes Store opened for business on 28 April 2003. On 24 February 2010, the store served its ten-billionth music download – a milestone reached in just under seven years – and currently accounts for seventy per cent of worldwide online digital music sales, making it the world's largest legal music retailer.[34]

From the point of view of this book, Jobs's most significant move came in 2007, when Apple entered the mobile phone market with a product – the iPhone – which has had a radically disruptive impact on the industry. It went on sale in July and was a typical Apple product – elegantly designed, beautifully made and incorporating an innovative touch-screen interface which displayed a virtual keyboard when required. But the most significant thing about the new phone was that it was essentially a powerful handheld computer that also had the capability to make voice calls. Of course all mobile phones are, to some extent, handheld computers, but up to then they had rather feeble processors, primitive operating systems and little Internet capability. The operating system of the iPhone, iOS, was a variant of OS X, the Unix-derived operating system which powered Apple's range of desktop and laptop computers. And it came with an email program and a version of Apple's Safari web browser. Subsequent versions of the iPhone added faster Internet connectivity, GPS, video capability and other features.[35]

The two most significant aspects of the new phone were its computing power and the fact that it gave users a taste of what it was like to have a workable Internet connection with them at all times.

Computing power: The processing power of the new device meant that it would be possible for developers to write serious programs called 'apps' (an abbreviation of 'applications') to run on it. In 2008 Apple released a Software Development Kit which enabled programmers to create applications for the phone. In July of that year Jobs launched the Apps Store as an integral part of the iTunes Store; this allowed iPhone users to download (and pay for) apps. Before they were allowed to appear in the store, however, all apps had to be pre-approved by Apple. Programs that were buggy or tasteless, or which threatened to affect the core functionality of the phone, were summarily rejected.[36] And the company took a thirty per cent cut of the price charged by the developer.

The result was the creation of a flourishing software ecosystem. Programmers had an incentive not only to create applications for an expanding market of affluent consumers who were accustomed to paying for stuff, but also had access to a global distribution system which incorporated a slick online payment system. Customers had ready access to a colossal range of inexpensive programs[37] that made their iPhones more enjoyable or useful. Most importantly, though, Apple remained in total control of the entire ecosystem. Nothing was sold on the iTunes Store without the company's explicit approval. And it went to great pains to ensure that users did not install 'unauthorized' software on their devices; if they did 'jailbreak' the device in this way, they ran the risk of having the phone disabled ('bricked' was the colloquial term) the next time they synchronized it with iTunes.

Thus the iPhone was what is known in the trade as a 'tethered' device: you might have thought that you owned your phone, but the reality was that it was controlled via a virtual wire that extended all the way back to Apple HQ in California. And unlike other brands of mobile phones, where the customer's relationship was with the mobile network that supplied connectivity, iPhone users alone continued to have a direct relationship with the company that manufactured their handsets. Apple's rationale for this extraordinary level of control was that it was the only way of ensuring that iPhones would be reliable, free of buggy or unscrupulous software and easy to synchronize and upgrade.

All of which was undoubtedly true. Steve Jobs provided his customers with a dependable, safe and high-quality product, in a way that earlier communications moguls like Sarnoff or Zukor would have recognized. But in opting for it, his customers were cheerfully – and voluntarily – queuing to enter what was, in effect, a luxuriously padded jail cell. And in the process they made his company the second most valuable public corporation in the world.

Mobile Internet connectivity: Prior to the arrival of the iPhone, mobile access to the Internet had been, to put it mildly, a pain. There were lots of reasons for this, ranging from the inadequate speeds and capacity of GSM networks, to the cluelessness of handset manufacturers, the computational feebleness of their devices, their primitive screens and user interfaces, and the rapacity of mobile network operators. But the end result was that if you wanted to have an acceptable user experience on the Internet then you had to be stationary and armed with a PC or a laptop.

The iPhone – especially the second version, the 3G model – changed that. The combination of an intuitive touch-screen interface, powerful processor and grown-up browser gave users a much better way of accessing the Net while on the move. A particularly innovative feature of the Apple device was the way it enabled users to enlarge or reduce a portion of a screen image by 'squeezing' or expanding the relevant part of the screen between thumb and finger. Web browsers running on earlier mobile handsets had been stuck with the problem of how to make web pages readable on small screens. Some had experimented with clunky 'zoom' facilities, but they had generally been unsatisfactory. The iPhone's combination of the Safari browser with the flexibility of its touch-screen interface showed what was possible.

And users loved it. The evidence was that iPhone owners discovered that they were using their laptops and desktop computers less. A lot less. And the reason was obvious: many of the purposes for which we access the Net are, computationally speaking, relatively trivial. We check email, read and post updates to social-networking sites, check timetables and calendars, read news headlines. For these kinds of tasks, a powerful computer with a fixed Internet connection is not really necessary. In the absence of a convenient mobile device we had no alternative but to use our computers for these purposes. But once a capable Internet-enabled phone arrived, that position changed.

The iPhone transformed the so-called 'smartphone' business, and rapidly achieved a significant market share – a development that took many observers by surprise. After all, when the iPhone launched in 2007 the mobile phone industry was a mature one, hitherto dominated by large companies like Nokia, Samsung, Sony Ericsson, RIM and Motorola. Apple had no experience in the industry, yet its product turned it upside down. While the incumbents thrashed around trying to find an answer to the challenge posed by the intruder, only one significant alternative emerged: Google's Android mobile operating system.

In some ways, the thinking behind Android paralleled that which went into the iPhone. It was based on the recognition that technological progress in chip design had made it possible to make phones that were essentially powerful handheld computers. But powerful processors need correspondingly sophisticated operating systems, and the problem for the incumbent handset manufacturers was that the software that ran their phones was relatively primitive. Worse, because the incumbents were primarily hardware manufacturers, they lacked the organizational competencies needed to create a really good operating system.

Google, on the other hand, *did* possess the requisite skills: it is, after all, primarily a software company. So it was able to develop the Android system as an open-source product derived from Linux, the free operating system which emerged as an alternative to the Unix system from which iOS was descended. But whereas Apple's strategy involved keeping tight control over the phone hardware as well as its software, Google's was to offer Android free to handset manufacturers, thereby enabling them to escape the consequences of their incompetence in software development.[38]

The outcome is that the smartphone market is increasingly dominated by two operating systems. On the one hand we have Apple, with the iPhone iOS and its associated software ecosystem; on the other we have a plethora of handset manufacturers offering phones running various versions of the Android operating system. The Apple product is expensive and tightly controlled; the Android alternative is cheaper and relatively open at the software level, but the handsets are largely controlled by the mobile operators which offer them on contract or pay-as-you-go tariffs.

The differences between the two sides of the smartphone business, however, should not blind us to the really significant thing that they have in common – the fact that they have created devices that make it practical to use the Internet while on the move. In that sense they have made the slogan 'The Internet in Your Pocket!' a reality. And the odds are that, one day in the not too distant future, most people will access the Net not from their laptops or PCs, but via their phones or similar mobile devices descended from them. And that's where the risk of the closure chronicled by Tim Wu comes back into focus.

The end of generativity?

A central argument of this book is that the Internet is best understood as a global system for springing surprises or – to put it more politely – a system for enabling disruptive innovation. One of the most convincing expositions of this idea is a book by the Harvard legal scholar Jonathan Zittrain with the paradoxical title *The Future of the Internet – and How to Stop It*.[39] Zittrain coins a term for the surprise-generation capability – *generativity*. The creativity and innovation we have seen over the last few decades, he argues, is the output of a generative system that has two components. One is the open Internet. The other is the programmable personal computer, the PC, via which people interact with the network.

In crude terms, the central argument of Zittrain's book is that if either component of the system becomes closed, then its astonishing generativity will be choked off or at any rate attenuated to the (slower) pace of innovation deemed acceptable by those who control it. As far as the network is concerned, his argument is that – much as I argue in Chapter 3 – the plague of malware that has been facilitated by the openness of the Internet's architecture will one day prove its undoing, either because of a catastrophic 9/11-type event, or because users who are worn down by the scourge of malware will eventually yearn for a safer, more controlled – and less innovative – information environment.

These pressures may also bear down on the other component of the generative system, the PC. In its classic form, a personal computer has an open architecture. That is to say, it will run any program that has been correctly coded for its operating system, no matter where that program originates from or who wrote it. And its owner can install on it any software that she or he desires.

Take for example the Apple computer on which I am writing this. It's a bog-standard machine with an Intel dual-core processor, some RAM, a screen and a couple of hard disks. At the moment, it is running OS X, the operating system that Apple supplies for it. But if I wish I can install Microsoft Windows 7 or Linux on it, and it will run them too. I can search the Internet for software written by companies and individuals, and download and install it if I wish. And my computer will run it, no questions asked.

If I'm a talented programmer (which, alas, I am not) I can write software for my machine. And I can distribute that software over the Net so that other people can run it on their computers. And at the moment there are no barriers to that process. That's generativity in action. It's how Shawn Fanning conceived and implemented Napster, the file-sharing system that so threatened the music industry. It's also how Tim Berners-Lee conceived, implemented and distributed the software for the Web.

But – Zittrain points out – if one of the programs I have downloaded contains some sinister, surreptitious code, then my computer will run that too. Or if I carelessly open an email attachment that has some embedded malware, or click on a website that likewise has some nasty code hidden in it, then my computer will run that code with possible nasty consequences for me or for somebody else. These are the downsides to openness, and they lead Zittrain to fret that eventually they may tempt many Internet users to move towards a device for accessing the Net that is less vulnerable to the downsides.

Enter Steve Jobs and his gorgeous, magical iPhone. 'We define everything that is on the phone,' he said as he announced the product in January 2007. 'You don't want your phone to be like a PC. The last thing you want is to have loaded three apps on your phone and then you go to make a call and it doesn't work anymore. These are more like iPods than they are like computers'.[40]

They are indeed. And so – to a lesser extent – are the Android-based and other smartphones. So one of the implications of the rise of the Internet-connected mobile phone is that sometime in the foreseeable future most people will access the Net not via an open, programmable device but via something that is functional, enjoyable and perhaps even beautiful – but is wholly or largely under someone else's control.

So what does our networked future hold?

'The future,' as George Soros, the financier and philanthropist once observed, 'is not only unknown: it is unknowable'. Given that, it seemed like a good idea to take Postman's two extremes to see how our networked future might unfold. Which kind of future – the Orwellian or the Huxleyean – seems more plausible? Postman thought that the two were mutually exclusive: humanity might have one or the other. But an even gloomier conclusion is that we might wind up with both.

On the Orwellian front, for example, things are well advanced – at least in terms of the technology needed for comprehensive surveillance. And although authoritarian regimes find it easy to deploy the technology to that end (they don't have to bother with seeking public consent), the liberal democracies are not far behind. In the latter case, the justification is the so-called 'war on terror', closely followed by the need to combat organized crime and paedophilia. Since 9/11, Western countries which pride themselves on adherence to the rule of law have passed increasingly intrusive legislation giving security, intelligence and law-enforcement agencies enhanced powers of Internet surveillance. Take, for example, the sweeping powers granted to government agencies by the PATRIOT Act that was rushed into law in October 2001. According to the Electronic Frontier Foundation,[41] the act greatly expands the four traditional tools of surveillance used by US law enforcement – wiretaps, search warrants, pen/trap orders and subpoenas – while reducing the checks and balances with which the US has traditionally limited state power. Among the new or expanded powers highlighted by the EFF are:

- The act allows the government to monitor the online activities of any American, and perhaps even track what websites s/he reads, by merely telling a judge anywhere in the US that the spying could lead to information that is 'relevant' to an ongoing criminal investigation. The person spied upon does not have to be the target of the investigation. This application must be granted and the government is not obligated to report to the court – or inform the person spied upon – what it has done.

- The law makes two changes to increase how much information the government may obtain about users from their ISPs or others who handle or store their online communications. Firstly, it allows ISPs voluntarily to hand over all 'non-content' information to law enforcement with no need for any court order or subpoena. Secondly, it expands the records that the government may seek with a simple subpoena (no court review required) to include records of session times and durations, temporarily assigned IP addresses, and means and source of payments, including credit card or bank account numbers.
- The act contains one new definition of terrorism, and three expansions of previous definitions also expand the scope of surveillance. Its definition of 'domestic terrorism' raises concerns about legitimate protest activity being prosecuted as terrorism, especially if violence erupts, while additions to three existing definitions of terrorism ('international terrorism', 'terrorism transcending national borders' and 'federal terrorism') expose more people to surveillance.

Similar kinds of laws now apply in most of the major Western democracies. In the UK, for example, there is the Regulation of Investigatory Powers Act (RIPA)[42] which became law in 2000. Among other things, RIPA:

- enables certain public bodies to demand that an ISP provides access to a customer's communications in secret
- enables mass surveillance of communications in transit
- enables certain public bodies to demand ISPs fit equipment to facilitate surveillance

- enables certain public bodies to demand that someone hand over keys to encrypted information
- allows certain public bodies to monitor people's Internet activities
- prevents the existence of interception warrants and any data collected with them from being revealed in court.

Most Internet users would be shocked if they realized the extent to which their online activities are already under surveillance by their governments. Indeed, I suspect that many are unaware of the laws which permit such surveillance. Given the comprehensiveness of the powers now granted by legislators to governments in this area, one might expect a significant amount of public concern. But there is little evidence of this; in fact, politicians and the media are far more exercised about suspected privacy violations by Internet companies like Google and Facebook than they are about anything that the police or intelligence services do in relation to our online activities. And most legislators seem blissfully ignorant or uninterested in the issue of online surveillance. When the RIPA Bill was before the UK parliament in 1999, for example, those of us who were seeking to have the bill amended were shocked to discover how apparently uninterested most MPs were in the issues it raised. My estimate at one stage was that of the 635 MPs in the House of Commons, only about 20 showed any interest in our concerns.[43] If the price of liberty in this area is eternal vigilance, then most UK politicians, not to mention the great British public, seem unwilling to pay it.

On the Huxleyean front, the question of whether our enjoyment of products and services will lead to the kind of industrial closure chronicled by Tim Wu and postulated by Jonathan Zittrain seems more open.

On the optimistic side, many well-informed people think that whole idea of 'closure' is absurd. In their eyes, the layered, decentralized technology of the Internet is just too powerful to be captured by any single organization or cartel. Wu points out that, in the end, it's governments that create monopolies – in the sense that they allow them to exist. The *Zeitgeist* of neo-liberal capitalism is hostile to overweening market power, as evidenced by the US Department of Justice's and the European Commission's anti-trust actions against Microsoft, and the European Commission's moves to investigate the market dominance of Google and Intel. These moves are in tune with citizens' suspicions of big business – a suspicion that was fuelled by the banking collapse of 2008 and the resultant government bail-outs. The conclusion of this line of thought is that the rise of a new corporate behemoth along the lines of the original AT&T is unlikely.

On the other hand . . .

Wu points out that some of the realities of power remain unchanged. Fifty years ago, communications power lay in control of the telephone line to the home (the so-called 'local loop') and of the electromagnetic spectrum. The former was controlled by AT&T in the US and by national telephone monopolies in other countries; the latter was controlled by governments, which were able to regulate what went on in the ether. And guess what? – it's still much the same today. So while the barriers to innovation on the open Internet are still low, they're very high when it comes to communications infrastructure, and that's what may determine how the future pans out.

Another compelling reality is the power of network effects – the phenomenon that the power and value of a network increases dramatically with the number of people who use it. A telephone network that has only a dozen subscribers has very little value, but one that has a million users is a different proposition. We've seen the power of network effects everywhere in the recent history of computing technology. It explains the rise and dominance of Microsoft, for example, which emerged from the chaos of incompatible PC operating systems of the late 1970s to produce what became the de-facto standard operating system for the industry – MS-DOS and, later, Windows. The more people who had Windows machines, the more profitable it became for developers to write software for the platform; the more software there was available, the more incentive there was for people to buy Windows machines. And so it went on.

We've seen the same phenomenon online – with many of the same results. Network effects explain, for example, how Google became such a dominant search engine, Facebook the dominant social-networking site, Amazon the leading online retailer, Apple the leading distributor of online music and eBay the dominant auction site. In each case, network effects push an organization to what begins to look like a 'natural' monopoly.

Finally, there's what one might call the Steve Jobs effect: the fact that consumers value products and services that are convenient, easy to use and classy – and that anyone who gives them these things will prosper. In that sense, perhaps Jobs was the true heir of the moguls who dominated the entertainment business in the twentieth century. 'This is what we figured out,' he once said. 'Americans want Hollywood content. They don't want Amateur Hour'. And who's to say that he was wrong?

So . . .?

The conclusion of this line of thinking is the thought that perhaps Postman was wrong: it's not an either-or scenario – not Orwell *or* Huxley, but *both*. This is the position that Evgeny Morozov takes in his thoughtful, brooding book, *The Net Delusion*.[44] To assume that all political regimes can be mapped somewhere on an Orwell-Huxley spectrum is, Morozov maintains, 'an open invitation to simplification; to assume that a government would be choosing between reading their citizens' mail or feeding them with cheap entertainment is to lose sight of the possibility that a smart regime may be doing both.'[45]

What's important to remember, though, is that these are just *scenarios* – that is to say, postulated sequences of events – rather than predictions. The truth is that nobody knows how the Internet is going to evolve, and any attempt to make detailed predictions about it is doomed to failure. The Orwell-Huxley dichotomy is useful solely because it illustrates the range of possibilities that we could be faced with as the network and society co-evolve. The unpalatable fact remains that the only sensible answer to questions about the future of the Internet is the one Chou en Lai gave all those years ago when asked for his opinion about the impact of the French Revolution: *it's too early to say*.

EPILOGUE

Someone once said that education is 'what you remember after you've forgotten everything you've been taught'. As a university teacher, it's an adage that I've always treasured, and so it was predictable that it would come to mind as this book draws to a close. What, I asked myself, do I want readers to remember long after they've forgotten the detail embedded in the nine preceding chapters?

Here's my answer.

Firstly, we need to take the long view. We're living through a revolutionary change in our information environment, and these are early days. We're in the same relative position as the good citizens of Mainz were in the 1470s, just two decades after Gutenberg launched his revolution: *they* had little idea of how things would pan out, and nor do we in relation to the Internet. That's no consolation to all those people who are demanding immediate answers to the problems and challenges posed by the Internet, but it's how things are. We need to get used to it.

Secondly, anyone hoping that the turbulence wrought by the Internet will eventually subside, and that things will eventually level out, is doomed to disappointment. The complexity of our emerging media ecosystem, together with the 'permissionless innovation' that is facilitated by the Internet, make a return to stability an unlikely prospect. Instead, our future will be one that is characterized by ongoing disruptive innovation. The good news is that we will adjust to this new reality, just as we have always done in the past. Humans are an adaptive species, and we are good at building tools that help us to cope with changing circumstances.

Thirdly, we need to rise above the optimist-pessimist, Utopian-dystopian dichotomies that characterize our current discussions about the Internet. Since the ancient Greeks we've known that technology giveth and technology taketh away. Plato fretted that the technology of writing would cause us to lose the art of remembering, just as my mathematics teacher worried that the arrival of slide-rules and calculators would mean that we would lose our ability to do mental arithmetic. Both were right: every technology-driven revolution involves gains and losses. That's why Joseph Schumpeter described the upheavals by which capitalism renews itself as waves of 'creative destruction'. Both words are important: the process is *creative*, in that great new possibilities are realized; but it's also *destructive* because things of real value are destroyed. The problem with the optimist-pessimist dichotomy is that the optimists rarely address the reality of destruction while the pessimists rarely acknowledge the creative possibilities of the new. We need to transcend this sterile shouting match. Like electricity, the networked information environment is here to stay; the technology cannot be un-invented, and our societies are already too dependent on it to drop it. It offers us great potential benefits, but the powerful vested interests that are threatened by it want to capture and control it for their own ends. And our dependence on it raises serious and worrying questions not just about security and privacy but also about cultural freedom, creativity and diversity. These are the things we should be talking about –and what this book is about.

And they are what I hope you will remember.

APPENDIX

A brief history of the Net

The Internet that we use today was switched on in January 1983. Its precursor was a military network, the Arpanet, which was constructed in the United States in the late 1960s, and which was essentially complete by the autumn of 1972. The Arpanet got its name from the Advanced Research Projects Agency (ARPA), a branch of the US Department of Defense which had been set up to foster advanced research in US universities and industrial laboratories in response to the Soviet Union's launch of Sputnik, the world's first artificial satellite, in 1957. Sputnik's launch, at the height of the Cold War, had triggered fears in the Eisenhower Administration that the US was falling behind its superpower rival in aerospace and other high-technology areas.

As part of its mission, ARPA funded a lot of research in computer science departments across the US. At the time this involved providing funds to enable research grantees to purchase hideously expensive mainframe computers. So ARPA wound up paying for dozens of these behemoths, all of which were incompatible with one another. They had different architectures, used different operating systems and had idiosyncratic communications interfaces. The result was a situation where an ARPA-funded researcher in one lab would effectively be unable to use an (ARPA-funded) mainframe in another laboratory without a good deal of tedious recoding, travel and other inconveniences.

These inconveniences were, in a way, the irritation that led to the Arpanet. In 1965, a young Texan named Bob Taylor became director of ARPA's Information Processing Techniques Office (IPTO) – the division of the agency responsible for advanced computing. Next to Taylor's office in the Pentagon was a small room labelled the 'Terminal Room'. In their history of the network, Katie Hafner and Matthew Lyon describe it thus:

> There, side by side, sat three computer terminals, each a different make, each connected to a separate mainframe computer running at three separate sites. There was a modified IBM Selectric typewriter terminal connected to a computer at the Massachusetts Institute of Technology in Cambridge. A Model 33 Teletype terminal, resembling a metal desk with a large noisy typewriter embedded in it, was linked to a computer at the University of California in Berkeley. And another Teletype terminal, a Model 35, was dedicated to a computer in Santa Monica, California, called, cryptically enough, the AN/FSQ 32XD1A, nicknamed the Q-32, a hulking machine built by IBM for the Strategic Air Command. Each of the terminals in Taylor's suite was an extension of a different computing environment – different programming languages, operating systems, and the like – within each of the distant mainframes. Each had a different log-in procedure; Taylor knew them all. But he found it irksome to have to remember which log-in procedure to use for which computer. And it was still more irksome, after he had logged

in, to be forced to remember which commands belonged to which computing environment.[1]

The incongruity of these arrangements was immediately obvious to Taylor. He was funding advanced research in a dispersed but increasingly effective community of computer scientists. But in order to communicate with that community he had to run the gauntlet of incompatible terminals and differing log-in procedures. It was, Hafner and Lyon wrote, like having a den cluttered with several television sets, each dedicated to a different channel. 'It became obvious,' Taylor told them, 'that we ought to find a way to connect all these different machines.'

The Arpanet

He found such a way. With an initial million dollars in funding, the IPTO awarded a contract to a Boston consulting firm, Bolt Beranek and Newman (BBN), to build the network that was christened the Arpanet. The designers started with a set of incompatible mainframe computers located at a number of remote locations.

There were basically two ways of getting them to communicate with one another.

The first was to program each one to speak a common communications language – a protocol, in computerspeak.

The disadvantage of this solution was that it would require an unconscionable amount of work. If there were N mainframes, then each would have to be programmed to communicate with N-1 incompatible peers. And the implicit combinatorial problem would grow as new, incompatible machines were added to the network.

The other solution was to interpose a sub-network of identical computers, each of which would act as an interface for its assigned mainframe.

The advantage of this solution was that it dramatically reduced the amount of software development required. Each mainframe would have to be programmed to communicate with only one machine – its 'Interface Message Processor' (IMP). Since all the IMPs were identical, they could easily communicate with one another. And this is the design that was adopted by the BBN engineers.

Work on the Arpanet project commenced in April 1969. The first two nodes of the network – a computer in UCLA and a machine at Stanford Research Institute – first exchanged messages in October 1969. By 1972 the network was essentially complete; it spanned the continental USA and even had a node across the Atlantic, in University College, London. The story of how it was built is an engrossing one that has been well told by Hafner and Lyon,[2] and so need not detain us here. But one thing is worth noting: it was accomplished in the teeth of scepticism and opposition from AT&T, the regulated monopoly that then operated the telephone network in the US.[3] This is an aspect of the story to which we will return.

The Internetworking project

The Arpanet was built with astonishing speed, but even in the short period of its construction the world hadn't stood still. In fact several other packet-switched networks had also been built in that time – though none on the scale of the Pentagon's network. In France, for example, there was the Cyclades network. In Britain a team at the National Physical Laboratory led by Donald Davies and Roger Scantlebury had built a fast, small-scale network. And another team at the University of Hawaii had constructed the Aloha network using radio links. Each of these networks had strong technical similarities to the Arpanet; all for example, used packet-switched technology. But they were owned and operated by independent agencies and were, to all intents and purposes, separate systems.

So when the ARPA engineers looked for the next challenge, it was, in a sense, staring them in the face: it was to invent a way of seamlessly integrating these autonomous networks plus the Arpanet into a global networks of networks – an 'internetwork'. Thus was born the 'internetworking' project. It started in 1973 and was led by two of the people who had worked most intensively on the Arpanet – Vinton (Vint) Cerf, and Robert Kahn.

The design problem and its solution

The problem that confronted the leaders of the internetworking project was simple to state but difficult to solve. It was simply this: how do you design a network that will link networks that you don't control and run applications that you cannot envisage or predict? Or, to put it another way, how do you avoid the problem exemplified by the telephone network, which was designed for one application (voice telephony) and proved totally unsuitable for a newer application (enabling computers to communicate with one another)?

The solution that Cerf, Kahn and their colleagues arrived at was beautifully simple, and can be expressed as two rules:

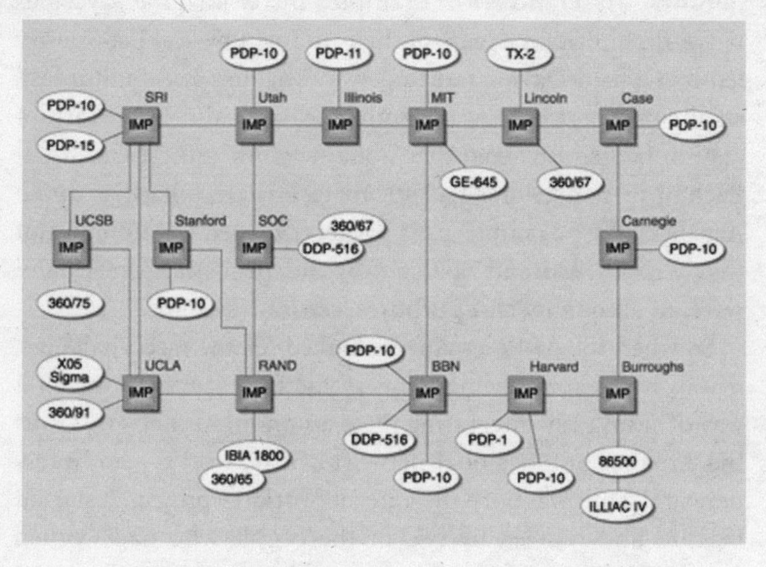

Figure A.1: The ARPAnet close to completion in 1971.

1. There should be no central control.
2. The network should be as simple as possible, and not optimized for any particular application (what became known as the 'end-to-end principle').

Rule 1 was necessary for two reasons. Firstly, it was an acceptance of reality – the agencies and states that ran the components of the 'internetwork' would never have agreed to central control exercised – in all probability – by an agency of the US government; and secondly, absence of central authority was necessary to avoid the kind of control exercised by AT&T over the US telephone network for most of the twentieth century – control which had limited the pace and scope of innovation to a pace determined by the Bell system.

Rule 2 resulted from the designers' recognition that they could not second-guess the future. So they headed down the road of 'dumb network, smart applications,' as one of them later put it. They came up with a design for a network that did only one thing: it took in data packets at one end and did its best to deliver them to their destinations. It left all the ingenuity to those at the ends of the network – the intellects who would use it. This was the 'end-to-end' principle.

Having settled on these two architectural principles, the Internetworking team devised a set of protocols (technical conventions) that would create a network that implemented the principles. The protocols enabled computers to act as gateways between different networks so that messages could be passed reliably from any node to any other node via an indeterminate number of routing stations. In the end, about a hundred separate protocols were required to define the operations of the network.

The two core protocols were:

- Transmission Control Protocol (TCP), which ensures that packets of data are shipped and received in the intended order
- Internet Protocol (IP), which specifies the format of packets and the addressing scheme

This pair of protocols is usually referred to as TCP/IP, and the dozens of supporting protocols are sometimes collectively referred to as the 'TCP/IP suite'.

So although the Internet is a formidably complex machine, it is based on two very simple principles, from which amazing developments flowed.

Dr Cerf's surprise-generation machine

The architectural principles underpinning the Internet's design had very far-reaching implications, some of which our societies are still struggling to comprehend.

No central control

The Cerf-Kahn design enabled the creation of a global network with an open, permissive architecture. There was no central control, and therefore it did not make sense to try to pre-specify the kinds of network that would be permitted to join. Anyone could hook up a network to the emerging 'internet-work' so long as they had a gateway computer that 'spoke' TCP/IP. This principle enabled the emerging network to grow organically at an astonishing speed. The absence of central control or ownership made the Internet radically different from previous communications networks, which had been owned or controlled by agencies that determined the uses to which their systems could be put.

In the UK and many European countries, for example, the national telephone networks were owned for most of the twentieth century by national post offices, which partly explains why fax technology was so slow to take off in the West: organizations devoted to delivering letters by hand were not disposed to take kindly to the idea of sending letters down a telephone wire. In sharp contrast, uses and applications of the Internet were determined entirely by the ingenuity of its users and those who developed applications that could harness its message-passing capabilities. The explosion of economic activity and creativity generated by the Internet is largely attributable to this factor.

The end-to-end principle

The second implication of the Cerf-Kahn design was that the overall network was essentially 'dumb'. Its only function was to pass data packets from one point to another – what became known as the 'end-to-end' principle. As far as the network was concerned, those packets might be fragments of email, photographs, recorded music or pornographic videos – they all looked the same to the network and were to be treated identically.[4] If you could do it with data packets, then the Internet would do it for you with no questions asked.

This indifference to content, combined with the permissiveness implicit in the absence of central control, meant that the network was designed to be relatively relaxed about who used it. Of course their computers had to 'speak TCP/IP', as it were, but if they did then nobody would deny them access to the network. This relatively laid-back philosophy meant that authentication (i.e. independent verification of identity) of users was not required. Each machine connected to the Internet needed to have a unique 'IP' (Internet Protocol) address, and all of that machine's transactions with other machines on the network could be logged. But there was – and currently still is – no provision for linking IP addresses to known individuals. This meant that the architecture facilitated anonymity – a feature famously encapsulated by the celebrated 1993 *New Yorker* cartoon showing two dogs in front of a computer. One is saying to the other, 'On the Internet, nobody knows you're a dog!'

The implications of the architecture's facilitation of

anonymity have been far-reaching. On the one hand it permits a wide range of reputable and disreputable uses of the Internet because the identity-based sanctions of the real world do not apply in cyberspace. On the other hand, anonymity enables the free expression and dissemination of views in ways that would be more difficult in real-world arenas. Anonymity makes it difficult, for example, for security services to track down or silence dissidents, and for corporations to identify whistle-blowers or campaigning groups disseminating critical information or hostile propaganda.

The technical architecture of the Internet has thus been a prime determinant of how the network came to be such an enabler of disruptive innovation.

ACKNOWLEDGEMENTS

I might not have embarked on this book had it not been for the insistence of two good friends, Nicci Gerrard and Sean French, that I should write it. Most advice of that kind is probably better resisted, but Sean and Nicci are such successful and accomplished writers that it was difficult to ignore them. So I didn't, and shall be eternally grateful to them for their persistence.

No writer is an island, as John Donne might have said. As an academic who works in two universities I'm blessed with colleagues who know more than I do, and I have shamelessly exploited my good fortune. At the Open University, Doug Clow, Glyn Martin, Ray Corrigan, Geoff Peters, Tony Hirst, Martin Weller and Kevin McConway resolutely ploughed through drafts and made penetrating comments and criticisms, some of which I found hard to answer – which is the way things ought to be. In Cambridge, Jon Crowcroft, Andrew Gruen, Quentin Stafford-Fraser, Huw Jones, Grant Young and Peter Morgan cast critical and helpful eyes over the text and suggested changes that, without exception, improved it. Jonathan Wardle of Bournemouth University took time out from a busy academic life to make some insightful observations, and my sons Brian and Peter looked at the draft from a friendly but critical perspective. My *Observer* colleague, Caspar Llewellyn-Smith, provided early encouragement by suggesting that I publish an outline of the book in the paper, thereby providing a firm deadline and corroborating Dr Johnson's observation that nothing concentrates a man's mind like the prospect of being hanged in the morning. I am grateful to him and to Jane Ferguson, the Editor of the *New Review*, for their confidence in a wayward columnist. I was gratified – and flattered – that friends like Andrew Ingram, Douglas McArthur, Duncan Thomas and David Hartley wanted to read a draft of the text. I'm grateful to my editor, Seán Costello, who went through the text with a perceptive and tactful eye, and to Richard Milner and Josh Ireland at Quercus who proved to be encouraging and understanding publishers. My greatest debt, though, is to Fiona Blackburn, who not only devoted endless hours to clearing away the undergrowth of my prose, but also supported its author while he was writing it.

Cambridge, July 2011

GLOSSARY

Algorithm: a sequence of instructions for carrying out a computational task.

API: an acronym for Application Programming Interface. Basically a set of rules and specifications that computer programs can follow to communicate with each other. An API serves as an interface between different programs and facilitates their interaction.

Application: a computer program designed to carry out a particular task or set of tasks. Examples include word processors, video players, spreadsheet programs.

Avatar: a graphical representation of the user (or of the user's alter ego or character) which can take either a 3-D form in games or virtual worlds like Second Life, or a 2-D form as an icon in Internet forums and other online communities.

Bandwidth: in digital communications, bandwidth is the measure, expressed in BITS per second, of the throughput of a data channel.

Bit: short for binary digit, i.e. one or zero.

Bitstreams: sequences of ones and zeroes in a digital communication.

Blog: a derivation of Weblog – an online diary in the form of a self-published website which generally consists of 'posts' (i.e. entries) presented in reverse-chronological order (i.e. more recent first).

Blogosphere: a collective noun used to describe the world of blogs, the links between them and their authors.

Bot: short for software robot – i.e. a computer program designed to carry out a particular task. The programs used by search engines to traverse the Web indexing web pages are examples of bots.

Botnet: a network of remotely controlled bots, generally covertly installed by **MALWARE** authors on unprotected Internet-connected computers, for nefarious purposes (e.g. **DDOS** attacks and relaying **SPAM**).

Broadband: a term used to describe communications channels with significant **BANDWIDTH**. Originally used to distinguish them from slower (lower bandwidth) telephone lines, but increasingly used to describe landline and wireless channels capable of handling 2 megabits per second or more.

Broadcasting: a term that originally came from agriculture, where it means to scatter seeds by hand rather than planting them in rows. In modern parlance, broadcasting usually means transmitting information via radio or a television. The key feature of broadcasting in this sense is that it is a few-to-many technology: a relatively small number of creators create content and transmit it to a large number of listeners or viewers.

Browser: a computer program which traverses information resources on the World Wide Web, and which retrieves and presents that information on a device's graphical display.

Clickstream: literally a sequence of mouse-clicks. In effect a record of the web pages one has accessed.

Client: in computing, a program (or, more generally, a system) that accesses a remote service on another computer system – a **SERVER** – via a network.

Cloud computing: a term that evolved to describe a system in which most of the computing resources needed to perform certain tasks are located somewhere on the Internet ('in the Cloud') rather than on users' PCs or portable devices. Common examples are web searching and web-mail like Gmail and Hotmail.

DDOS: an acronym for Distributed Denial of Service attacks in which computers in a **BOTNET** bombard a targeted **SERVER** with requests until it is overwhelmed.

DNS: acronym for Domain Name System. This is the hierarchical naming system which translates domain names meaningful to humans (e.g. www.bbc.co.uk) into the numerical IP address of another computer (in this case a BBC server identified as 212.58.244.68) for the purpose of locating and addressing Internet-connected devices worldwide.

Feed reader: a computer program that automatically visits sites to whose RSS feeds a user subscribes and downloads the resulting stream of snippets for reading later.

HTML: acronym for Hypertext Mark-up Language – a set of conventions for tagging the content of web pages so that a **BROWSER** can render them for viewing on a user's device.

HTTP: acronym for Hypertext Transfer Protocol – a networking **PROTOCOL** for distributed hypermedia information systems like the World Wide Web. HTTP regulates the interactions between a web **BROWSER** and a web **SERVER** as the former requests – and receives – pages from the latter.

Hyperlink: a reference in a web page to another web page or document that the reader can directly follow by clicking on it.

Hypertext: literally, non-linear text. Generally used nowadays to refer to text displayed on a computer or other electronic device with **HYPERLINKS** to other documents that the reader can immediately access, usually by click on a mouse or pressing a key.

Internet Protocol: one of the Internet's fundamental **PROTOCOLS**. It defines the way individual computers on the network are identified, and how data packets are routed through intermediate points on their path through the network.

Internet Relay Chat: a form of real-time text messaging or **SYNCHRONOUS** conferencing which is mainly used for group communication over the Internet.

Internet Service Provider: a company or other organization (e.g. a university) that provides Internet connections for customers or subscribers.

IP: acronym for **INTERNET PROTOCOL**.

IP address: the unique address of a computer that is connected to the Internet. In the original addressing system designed for the Internet (known as IPv4), IP addresses are expressed as a set of four three-digit numbers separated by dots (e.g. 174.112.255.113). The number of possible unique addresses within this system (over four billion) has proved insufficient to keep up with the growth of the network, and so an extended addressing system (IPv6) is slowly being phased in. Since numerical IP addresses are not exactly user-friendly, they are translated by the **DNS** system which maps 'real' names (e.g. www.bbc.co.uk) onto a numerical address.

IRC: acronym for **INTERNET RELAY CHAT**.

ISP: acronym for **INTERNET SERVICE PROVIDER**.

Malware: a collective term for malicious software like computer viruses, worms or Trojans together with **SPAM**.

Mashup: a web page or application that combines data or functionality from two or more sources to create new services, generally using **APIs**. The idea behind a mashup is to produce enriched results (e.g. graphical visualizations overlaid on maps) that were not necessarily envisaged by the producers of the data sources that they combine.

Meme: an infectious idea which replicates itself. Analogous to the biological concept of a gene.

Multi-tasking: a way of operating a computer in such a way that the machine appears to be performing a number of different tasks simultaneously. (In fact it is cycling rapidly between each task, but if this is done quickly enough the illusion of simultaneity is preserved.)

Open Source: a term used to describe computer software that is made available in human-readable format under licences that permit users to modify the code in ways that suit their purposes and to make the modified code available under the same terms as they received it.

Operating system: a computer program that manages the various hardware resources of a particular machine (e.g. disks, RAM, etc.) and provides common services for execution of **APPLICATION** software. Examples include Microsoft Windows, Apple OS X and Linux. The operating system is the most important program on a computer system. Without it, the machine is basically just an expensive paperweight.

Peer-to-peer networking: a technology (sometimes referred to as P2P) which enables two Internet-connected computers to connect directly with one another for the purpose of exchanging data.

Protocol: an agreed set of procedures to be followed when two computers or computing devices communicate. The **HTTP** protocol that governs interactions between a web **BROWSER** and a **SERVER** is an example of a protocol.

RSS: acronym for Really Simple Syndication (and sometimes also called Rich Site Syndication). It refers to a group of formats used to publish frequently updated works (e.g. blog entries and news headlines) in a standardized format. An RSS document is a simple text document (called a 'feed' or a 'channel') that includes summarized (or full) versions of the updated web material, plus 'metadata' such as time and date of publication, authorship, etc.

Server: can refer to (1) a computer program which serves the requests of other programs (**CLIENTs**) which may or may not be running on the same computer; (2) a physical computer dedicated to servicing the needs of **CLIENT** programs running on other computers on the same network; or (3) a 'software+hardware' system (i.e. a program running on a dedicated computer) which handles access to database, files, email or printing services.

SMTP: acronym for Simple Mail Transfer Protocol. The **PROTOCOL** that handles electronic mail carried over the Internet.

Social graph: a diagrammatic way of representing the online relationships between members of an online community.

Spam: unsolicited ('junk') email.

Streaming media: a technology for transmitting audio or video over the Internet. Differs from downloading because the recipient doesn't have to wait for the entire file to be transmitted before s/he can view or listen. After a brief pause, the stream can be listened to, or viewed.

Synchronous: happening in real time. Instant Messaging,

Internet Chat and Skype conferencing are examples of synchronous communication. Email, in contrast, is an asynchronous system, because sender and recipient don't have to be online at the same time.

Thermionic valve: a glass-enclosed device that operates by the flow of electric current through a vacuum. These valves were critical to the development of the early electronic industry because they enabled signals to be amplified (e.g. in radios and public-address systems) or switched (as in early digital computers).

UNIX: A type of **OPERATING SYSTEM** developed at Bell Labs in the US in the 1960s which has been very influential in the evolution of computing.

URI: acronym for Uniform Resource Indicator, a string of characters used to identify a resource on the Internet.

URL: acronym for Uniform Resource Locator. Often interpreted as the 'web address' of a particular page, but in fact it can do more because it also specifies the method of accessing the resource. Web URLs are prefaced by **HTTP** – which specifies the **PROTOCOL** to be used when accessing the resource. But other protocols (e.g. FTP – for 'File Transfer Protocol') could also be used.

VoIP: stands for 'Voice over Internet Protocol'. In everyday terms it's using the Internet to convey voice conversations. It works by converting the pressure waves created by the speaker's voice into data packets, which are then sent over the network in the usual manner. At the destination end, the packets are disassembled and converted back into audio signals. Skype is an example of a VoIP system.

Usenet: a global, Internet-based system of discussion groups which was very influential in the early days of the Internet.

Wiki: a user-editable web page. Originally, web pages were read-only publications. Wiki technology enabled Web users to edit pages and then have the edited version instantly or quickly republished. Wiki technology is what powers Wikipedia.

Zero-sum game: a game in which there is a winner and a loser. More precisely, a game in which a participant's gain or loss is exactly balanced by another participant's loss or gain.

NOTES

Prologue: Why this book?

1. John Naughton, *A Brief History of the Future: The Origins of the Internet*, Phoenix, 2000.
2. Manuel Castells, *The Internet Galaxy*, Oxford.
3. *Psychological Review*, 1956, vol. 63, pp. 81–97.
4. Probably a reference to Senator Joseph McCarthy.
5. Defined by the indispensable Jargon File as 'Large, expensive, ultra-fast computers. Used generally of number-crunching supercomputers, but can include more conventional big commercial IBMish mainframes. Term of approval; compare heavy metal, oppose dinosaur.' See http://www.catb.org/jargon/html/B/big-iron.html (accessed 14 February 2011).
6. John Naughton, 'The 9 things you need to know about the Internet', *Observer* 20 June 2010. Online at http://memex.naughtons.org/archives/2010/06/20/11251 (accessed 14 February 2011).
7. Although one – who is one of the most knowledgeable geeks I am privileged to know – described the piece as a

'fantastic read and a marvel of economy, managing to pack nine very big ideas into 15 minutes' reading. This is the kind of primer you want to slide under your boss's door.' http://www.boingboing.net/2010/06/20/nine-things-you-need.html (accessed 14 February 2011).

1. Take the long view

1. The phrase is attributed to John Gilmore, a famous engineer and activist who was one of the founders of the Electronic Frontier Foundation. See http://en.wikipedia.org/wiki/John_Gilmore_(activist) (accessed 13 June 2011).
2. This was the title of a book by Thomas Friedman, a well-known columnist on the *New York Times*.
3. John Man's book *The Gutenberg Revolution: How Printing Changed the Course of history*, Headline, 2002 provides a very readable account of what's known about Gutenberg's life. And Blake Morrison's *The Justification of Johannes Gutenberg*, Vintage, 2001 is a clever fictionalized account of Gutenberg's life, as seen by himself.
4. http://www.bl.uk/treasures/gutenberg/basics.html
5. Elizabeth Eisenstein, *The Printing Revolution in Early Modern Europe*, Cambridge University Press, 1993, p.15.
6. Neil Postman, *The Disappearance of Childhood*, Vintage Books, 1996, p.25.
7. http://www.nielsenbookdata.co.uk/uploads/press/1Nielsen Book_ProductionFigures_Feb09.pdf

8. Myron Gilmore, *The World of Humanism*, Harper & Brothers, 1952, p.186.
9. Adrian Johns, *The Nature of the Book: Print and Knowledge in the Making*, University of Chicago Press, 2000.
10. Postman, *op. cit.*, pp.21–2.
11. George Sarton, *The Renaissance: Six Essays*, Indiana University Press, 1962, p. 66.
12. Eisenstein, op. cit. , p.148.
13. See http://en.wikipedia.org/wiki/Second_Vatican_Council (accessed 17 April 2011).
14. Thomas Kuhn, *The Structure of Scientific Revolutions*, University of Chicago Press,1962.
15. Postman, *op. cit.*, p.18.
16. Postman, *op. cit.*, pp.21–2.
17. Postman, *op.cit*, p.26.
18. See Sarah Bakewell, *How to Live: A Life of Montaigne in One Question and Twenty Attempts at an Answer*, Chatto & Windus, 2010.
19. Norman Mailer, *Advertisements for Myself*, Andre Deutsch, 1961.
20. Andrew Sullivan, 'Why I Blog', *Atlantic*, November 2008. Online at http://www.theatlantic.com/magazine/archive/2008/11/why-i-blog/7060/.
21. Postman, *op. cit.*, p.27.
22. Postman, *op. cit*
23. Maryanne Wolf, *Proust and the Squid: The Story and Science of the Reading Brain*, Icon Books, 2008.

24. Wolf, *op. cit.* p.5.
25. Postman, *op. cit.*, p.51.
26. Postman, *Amusing Ourselves to Death*, Methuen, 1987, p. 51.
27. William Gibson, *Neuromancer*, Gollancz, 1984.
28. Bruce Sterling, *The Hacker Crackdown*, Spectra Books, 1992.
29. Jack Goldsmith and Timothy Wu, *Who Controls the Internet? Illusions of a Borderless World*, Oxford University Press, 2006, p.13.
30. Goldsmith and Wu, *op. cit.*, p.27.
31. http://w2.eff.org/Censorship/Internet_censorship_bills/barlow_0296.declaration (accessed 28 March 2010).
32. The CDA punished all transmissions of 'indecent' sexual communications or images on the Internet 'in a manner available to a person under 18 years of age'. The act was challenged by the Electronic Frontier Foundation, a campaigning organization founded by Barlow and others, and eventually declared unconstitutional by the US Supreme Court in June 1997.
33. See, for example, the extraordinary speech on 'Internet Freedom' delivered by the US Secretary of State, Hilary Clinton, on 21 January 2010. The full text is available at http://www.foreignpolicy.com/articles/2010/01/21/internet_freedom?page=full (accessed 27 March 2010).
34. See James Surowiecki, *The Wisdom of Crowds: Why the Many Are Smarter Than the Few*, Abacus, 2005.

35. See Steve Weber, *The Success of Open Source*, Harvard, 2005.
36. See Yochai Benkler, *The Wealth of Networks*, Yale, 2006.
37. But for an admirable attempt see Michiko Kakutani, 'Texts Without Context', *New York Times*, 17 March 2010. Available online at http://www.nytimes.com/2010/03/21/books/21mash.html (accessed 30 March 2010).
38. Available online at http://www.theatlantic.com/magazine/archive/2008/07/is-google-making-us-stupid/6868/. Coincidentally, the *Atlantic* is the magazine which in 1945 published Vannevar Bush's prophetic essay 'As We May Think' which many people view as an outline sketch for the Web. See John Naughton, *A Brief History of the Future: The Origins of the Internet*, Phoenix, 2000 pp. 213–16.
39. Bruce Friedman, 'How Google is changing our information-seeking behaviour', Blog post dated 6 February, 2008. Available online at http://labsoftnews.typepad.com/lab_soft_news/2008/02/how-google-is-c.html.
40. Quoted in Carr, *op. cit.*
41. Nicholas Carr, *The Shallows*, forthcoming from W.W. Norton at the time of writing. See http://www.theshallowsbook.com/nicholascarr/The_Shallows.html (accessed 14 April 2010.)
42. See, for example, the discussions organized by Edge.org at http://www.edge.org/discourse/carr_google.html (accessed 31 March 2010) and by the Encyclopaedia

Britannica Blog at http://www.britannica.com/blogs/2008/07/this-is-your-brain-this-is-your-brain-on-the-internetthe-nick-carr-thesis/ (accessed 31 March 2010).

43. Jamais Cascio, 'Get Smarter', *Atlantic*, July/August 2009. http://www.theatlantic.com/magazine/print/2009/07/get-smarter/7548/ (accessed 31 March 2010)
44. Cascio, *op.cit.*
45. Douglas Engelbart is a pioneering computer scientist who invented many of the key features of modern graphics-based computer interfaces, including the mouse, windows and bit-mapped graphics screens. For an account of his contribution see John Naughton, *A Brief History of the Future: The Origins of the Internet*, Phoenix, 2000, pp.216–20.
46. *New York Times*, 14 October 2009. Available online at http://roomfordebate.blogs.nytimes.com/2009/10/14/does-the-brain-like-e-books/ (last accessed 14 April 2010).
47. http://www.vetmed.wsu.edu/research_vcapp/Panksepp/ (accessed 13 June 2011).
48 Emily Yoffe, 'Seeking: How the brain hard-wires us to love Google, Twitter, and texting. And why that's dangerous', *Slate*, 12 August 2009. Available online at: http://www.slate.com/id/2224932/pagenum/all/#p2 (accessed 14 April 2010).
49. Clay Shirky, 'Tools and Transformations', PenguinBlog USA, 3 November 2008. http://us.perguingroup.com/static/html/blogs/tools-and-transformations-clay-shirky (accessed 13 April 2010).

2. The Web is not the Net

1. You think I jest? The 2009 annual conference of the UK Society of Editors in Stansted was addressed by Helen Boaden, the Director of BBC News. In the course of her remarks on 16 November she referred to Tim Berners-Lee as 'the man credited with inventing the Internet'. Berners-Lee is, of course, the inventor of the World Wide Web, and it was probably a slip of the tongue by the usually well-informed Boaden, but it illustrates my point nicely.

3. For the Net, disruption is a feature, not a bug

1. "The Net Imperative", *The Economist* , 24 June, 1999.
2. http://www.claytonchristensen.com/disruptive_innovation.html (accessed 17 April 2011).
3. Lawrence Lessig, *The Future of Ideas: The Fate of the Commons in a Connected World*, Vintage, 2002, p.44.
4. Tim Wu, *The Master Switch: The Rise and Fall of Information Empires*, Atlantic Books, 2011, p.106.
5. For a detailed account of the network's architecture, see Barbara van Schewick, *Internet Architecture and Innovation*, MIT Press, 2012.
6. Tim Berners-Lee, *Weaving the Web*, Oxford, 1999, p.17.
7. Berners-Lee, *op. cit.*, p.28.
8. The relevant technical paper is http://www.faqs.org/rfcs/rfc1630.html (accessed 17 April 2011).
9. A British student on an internship, Nicola Pellew, wrote a simple, line-mode browser that could be run on

computers that were less exotic than the NeXT.

10. Berners-Lee, *op. cit.*, p.33.
11. Berners-Lee, *op. cit.*, p.35. In the same passage he notes that this policy of porting the browser to many different computers involved making a strategic compromise. He had originally intended that the browser should enable its user not only to view web pages, but also to create and edit them. In other words, he originally envisaged the Web as a read-write medium rather than a read-only one. It was only when Web 2.0 applications arrived that this dream began to be realized.
12. Berners-Lee, *op.cit.*, p.51.
13. See http://www.worldwidewebsize.com/
14. Eric S. Raymond: *The Cathedral and the Bazaar: Musings on Linux and Open Source by an Accidental Revolutionary*, O'Reilly, 1999, p.32.
15. John Alderman, *Sonic Boom: Napster, P2P and the Future of Music*, 4th Estate, 2002.
16. In his book *Being Digital*, Knopf, 1995.
17. Which, coincidentally, was the academic field in which J.C.R. Licklider, widely regarded as the man who inspired the ARPAnet project, specialized. See M. Mitchell Waldrop, *The Dream Machine: J.C.R. Licklider and the Revolution That Made Computing Personal*, Viking, 2001.
18. The specification is available from http://www.iso.org/iso/catalogue_detail.htm?csnumber=22412 – for a fee.
19. See http://www.mp3licensing.com/mp3/index.html. This

means that MP3 encoding technology is not free but proprietary, so creators of encoders that use it must obtain a licence from Fraunhofer's agents. Fraunhofer obtained a German patent on MP3 in 1989 and a US patent in 1996. See http://inventors.about.com/od/mstartinventions/a/MPThree.htm for details.

20. Which Apple, the computer company, later appropriated as a sales slogan for its iTunes/iPod combination of music software and hardware.
21. As we shall see, the original Napster founded by Shawn Fanning was eventually shut down by legal action. A new company, also called Napster but providing legal access to streamed and downloaded music, was established in 2003. It can be found at www.napster.com and www.napster.co.uk. All reference to Napster in this chapter relate to the original enterprise.
22. Paul Goldstein, *Copyright's Highway: From Gutenberg to the Celestial Jukebox*, Stanford University Press, 2003.
23. http://en.wikipedia.org/wiki/Napster#Shutdown (accessed 14 June 2011).
24. Jonathan Zittrain, *The Future of the Internet – and How to Stop It*, Yale University Press, 2008, p.36.
25. Morris was quickly identified as the author of the worm, which he had launched by logging into an MIT machine to conceal the fact that he was actually based in Cornell. He was tried and convicted of violating the 1986 Computer Fraud and Abuse Act. After appeals he was sentenced to 3 years probation, 400 hours of community service, and a fine of $10,000, but the incident seems not

to have had any long-term deleterious impact on his career. He moved from Cornell to Harvard, founded a dot-com company with some friends which they sold to Yahoo! in 1998 for $49 million, and now has a tenured position at MIT.

26. Most of today's computer users have never used – or seen – a floppy disk; today's computers boot from hard drives or solid-state devices.
27. A networking technology known affectionately as 'sneakernet'.
28. Zittrain, *op. cit.*, p.44.
29. *ibid*
30. It's not clear whether the acronym coined by Hormel Foods, maker of the canned food product, derived from 'Shoulder Pork and hAM' or 'SPiced hAM'. Either way, it's SPAM.
31. DEC is now defunct, though formally the property of Hewlett-Packard, which acquired it when it bought Compaq. One of the reasons it went under was that its founder, Ken Olsen, failed to spot the threat posed by the PC to his company's product line.
32. For the background and full text see http://www.templetons.com/brad/spam/spam25.html
33. http://www.templetons.com/brad/spamterm.html. The sketch is available on YouTube at http://www.youtube.com/watch?v=anwy2MPT5RE (accessed 3 October 2011).
34. http://www.viruslist.com/en/analysis?pubid=204792072 (accessed 4 January 2010).

35. http://www.theregister.co.uk/2003/11/18/the_economics_of_spam/ (accessed 4 January 2010).
36. http://www.schneier.com/blog/archives/2008/11/the_economics_o.html (accessed 4 January 2010).
37. http://www.eurocomms.com/features/7161-7161 (accessed 8 April 2011).
38. Available online at *www.official-documents.gov.uk/document/cm76/7650/7650.pdf*
39. http://www.oecd.org/document/54/0,3746,en_2649_34225_38690102_1_1_1_1,00.html (accessed 3 October, 2011).
40. Variously called a 'bot herder' or even a 'botmaster'.
41. http://en.wikipedia.org/wiki/Conficker (accessed 18 January 2010).
42. The 'Guide' was visualized as an electronic encyclopedia, compiled using an innovative editorial formula. 'Most of the actual work got done by any passing stranger who happened to wander into the empty offices of an afternoon and saw something worth doing'. Douglas Adams, *Life, the Universe and Everything* in *The Hitchhiker's Guide to the Galaxy: A Trilogy in Four Parts*, Pan, 1992.
43. Data from Alexa.com: http://www.alexa.com/topsites (accessed 17 April 2011).
44. http://www.quora.com/How-many-employees-does-Wikipedia-have (accessed 17 April 2011).
45. Aniket Kittur, Ed H. Chi and Bongwon Suh, 'What's in Wikipedia? Mapping Topics and Conflict Using Socially Annotated Category Structure', *Proceedings of the 27th International Conference on Human Factors in Computing*

Systems (Boston, MA, USA, April 4–9 2009), pp.1509–12.

46. Stacy Schiff, 'Know It All', *New Yorker*, 31 July 2006. Online at: http://www.newyorker.com/archive/2006/07/31/060731fa_fact (accessed 17 April 2011).
47. James Boyle, 'Cultural Agoraphobia and the Future of the Library', First Arcadia Lecture, Cambridge University, 12 March 2009.
48. Wikipedia, 'Wikipedia Biography Controversy', http://en.wikipedia.org/wiki/Seigenthaler_incident (accessed 17 April 2011).
49. Jim Giles, 'Internet encyclopedias go head to head', *Nature* 438, 15 December 2005, pp.900–01.
50. Andrew Dalby, *The World and Wikipedia: How We Are Editing Reality*, Siduri Books, 2009, p.8.
51. Clay Shirky, *Cognitive Surplus: Creativity and Generosity in a Connected Age*, Allen Lane, 2010, p.9.
52. Yochai Benkler, *The Wealth of Networks: How Social Production Transforms Markets and Freedom*, Yale, 2006, p.122.
53. For more about Cunningham's work see Chapter 7.
54 http://en.wikipedia.org/wiki/Facebook (accessed 17 April 2011).
55. http://www.quora.com/How-many-photos-are-uploaded-to-Facebook-each-day (accessed 17 April 2011).
56. 'Social Network Sites: Definition, History and Scholarship', Boyd, D. M., & Ellison, N. B. (2007). 'Social network sites: Definition, history, and scholarship', *Journal of Computer-Mediated Communication*, 13(1),

article 11. Available online at http://jcmc.indiana.edu/vol13/issue1/boyd.ellison.html (accessed 3 October, 2011)

57. *ibid.*
58. http://www.burningman.com/whatisburningman/ (accessed 17 April 2011).
59. http://techcrunch.com/2011/03/23/amazingly-myspaces-decline-is-accelerating/ (accessed 17 April 2011).
60. Arrow, 2010.
61. Bari M. Schwartz, 'Hot or Not? Website Briefly Judges Looks', *Harvard Crimson*, 4 November 2003, http://www.thecrimson.com/article/2003/11/4/hot-or-not-website-briefly-judges/ (accessed 23 November 2010).
62. Code for White Anglo-Saxon Protestant.
63. Lawrence Lessig, 'Sorkin vs. Zuckerberg', *New Republic*, 1 October 2010. Online at http://www.tnr.com/article/books-and-arts/78081/sorkin-zuckerberg-the-social-network (accessed 11 February 2011).
64. *ibid.*
65. The estimated valuation of Facebook in June 2011.
66. Boyd and Ellison, *op.cit.*, p.1.
67. van Schewick, *op. cit.*, p.207.
68. *ibid.*

4. Think ecology, not just economics

1. See, for example, John Cassidy: 'After the Blowup', *New Yorker*, 11 January 2010, p.28.

2. See, for example, Carl Shapiro and Hal R. Varian, *Information Rules: A Strategic Guide to the Network Economy*, Harvard Business School Press, 1998. Varian is now Chief Economist at Google.
3. Yochai Benkler, *The Wealth of Networks: How Social Production Transforms Markets and Freedom*, Yale, 2006.
4. What the cultural critic Clay Shirky calls 'the mass amateurization of publishing'. See http://www.shirky.com/writings/Weblogs_publishing.html (accessed 17 April 2011).
5. As I write, about 48 hours'-worth of video material is uploaded to YouTube every *minute*. See http://allthingsd.com/20110525/youtube-is-6-years-old-and-huge-3-billion-views-a-day/ (accessed 15 June 2011).
6. See Steven Weber, *The Success of Open Source*, Harvard University Press, 2004.
7. Benkler, *op. cit.* p. 3.
8. As distinct from copyrighted content that is 'free' only because those re-publishing it have not paid royalties on it.
9. http://www.worldwildlife.org/climate/curriculum/item5957.html
10. Martin Weller, *Virtual Learning Environments: Using, Choosing and Developing Your VLE*, Routledge, 2007.
11. Bonnie A. Nardi and Vicky L. O'Day, *Information Ecologies: Using Technology with Heart*, MIT Press, 1999, p. 53.
12. John Seely Brown and Andrew Duguid, *The Social Life of Information*, Harvard Business School Press, 2000.

13. '"Big Media" Meets the "Bloggers": Coverage of Trent Lott's Remarks at Strom Thurmond's Birthday Party', Kennedy School of Government, Harvard University, Case Program, 2004.
14. Jay Rosen, 'The Legend of Trent Lott and the Weblogs', *PressThink*, 15 March 2004. Available online at: http://blogcritics.org/culture/article/the-legend-of-trent-lott-and/ (accessed 3 October 2011).
15. Michael Dobbs and Mike Allen (10 September 2004). 'Some Question Authenticity of Papers on Bush'. *Washington Post.* p.A01. http://www.washingtonpost.com/wp-dyn/articles/A9967-2004Sep9.html (accessed 3 October 2011).
16. Rather has denied that it was the *60 Minutes* episode that brought his career to a premature end. 'There were plans before the Bush story ever appeared,' he said in an interview, 'for me to leave the Evening News at or about the time that I did. I'd been there 24 years, and I would have liked to have made 25, but I knew the preceding summer that that wasn't going to happen. They said they wanted to take CBS News in different directions, and consequent events have proven that they did want to take it in different directions. But promises were made and not kept. Contracts that had been written were not adhered to, and I didn't feel terrific about that.' http://www.pbs.org/wgbh/pages/frontline/newswar/tags/rathergate.html
17. http://en.wikipedia.org/wiki/Wikileaks (accessed 28 March 2011).
18. For this case-study I have drawn heavily on a pre-print

version of Yochai Benkler's paper, 'A Free Irresponsible Press', which at the time of writing was forthcoming from *Harvard Civil Rights-Civil Liberties Law Review*. Available online from http://www.benkler.org/Benkler_Wikileaks_current.pdf (accessed 12 April 2011).

19. *ibid.*
20. http://wikileaks.ch/About.html (accessed 12 April 2011).
21. Benkler, *op. cit.*, p.64.
22. Benkler, *op. cit.*, p.18.
23. See http://www.guardian.co.uk/world/2011/sep/01/unredacted-us-embassy-cables-online (accessed 2 October, 2011).
24. *Washington Post*, 30 November 2010. Online at: http://www.washingtonpost.com/wp-dyn/content/article/2010/11/30/AR2010113001095.html (accessed 13 April 2011).
25. http://www.huffingtonpost.com/2010/12/19/joe-biden-wikileaks-assange-high-tech-terrorist_n_798838.html (accessed 13 April 2011).
26. Thomas L. Friedman, 'We've Only Got America A', *New York Times*, 14 December 2010. Online at: http://www.nytimes.com/2010/12/15/opinion/15friedman.html (accessed 13 April 2011).
27. Ed Pilkington, 'Bradley Manning: top US legal scholars voice outrage at "torture"', *Guardian*, 10 April 2011. Online at: http://www.guardian.co.uk/world/2011/apr/10/bradley-manning-legal-scholars-letter (accessed 13 April 2011).

28. 'WikiLeaks Founder on the Run, Trailed by Notoriety', *New York Times*, 23 October 2010. Online at: http://www.nytimes.com/2010/10/24/world/24assange.html (accessed 13 April 2011).
29. Bill Keller, 'The Boy Who Kicked the Hornet's Nest', in Alexander Star (editor), *Open Secrets: WikiLeaks, War and American Diplomacy*, Grove Press, 2011.
30. *ibid.*
31. Dan Gillmor, *We the Media: Grassroots Journalism by the People, for the People*, O'Reilly, 2006, p.187.
32. http://www.pbs.org/wgbh/pages/frontline/newswar/interviews/rosen.html
33. See Tom Standage, *The Victorian Internet*, Phoenix, 1999.
34. i.e. few-to-many.
35. Newspaper evangelists were fond of pointing out that there was more information – or at any rate more words – in a single page of *The Times* than in an entire 30-minute TV news bulletin.
36. See Philip Evans and Thomas S Wurster, *Blown to Bits: How the New Economics of Information Transforms Strategy*, Harvard Business School Press, 1999.
37. See www.craigslist.com/about/sites (Last accessed 9 May 2010.)
38. http://www.technewsworld.com/story/39313.html?wlc=1250798432&wlc=1273413854 (Last accessed 9 May 2010).
39. As observed by, for example, the cultural critic Neil Postman in his book *Amusing Ourselves to Death* and by

the journalist James Fallows in *Breaking the News: How the Media Undermine American Democracy*, Pantheon Books, 1996.

40. Skinner, B. F., '"Superstition" in the Pigeon', *Journal of Experimental Psychology*, 38(2), 1948, pp.168–72.
41. Because transmissions on adjacent frequencies interfere with one another if the gap between them is too narrow.
42. One indicator that Internet-vocal customers are having an impact is the growing trend – at least in the US – for corporations to resort to legal threats to silence dissatisfied online customers. See Dan Frosch, 'Venting Online, Consumers Can Find Themselves in Court', *New York Times*, 31 May 2010. Online at http://www.nytimes.com/2010/06/01/us/01slapp.html (accessed 2 June 2010).
43. http://blog.flickr.net/en/2010/09/19/5000000000/ (accessed 8 April 2011).
44. See creativecommons.org.
45. See 'Viacom Sues Google Over YouTube Video Clips', New York Times, 14 March, 2007. Online at http://www.nytimes.com/2007/03/14/business/14viacom.web.html (accessed 3 October, 2011).
46. http://www.youtube.com/watch?v=_OBlgSz8sSM (accessed 24 May 2010, at which date it had been viewed 193,562,789 times).
47. Lawrence Lessig, *Remix: Making Art and Culture Thrive in the Hybrid Economy*, Bloomsbury Academic, 2008.
48. http://www.imdb.com/title/tt0363163/

49. http://www.newyorker.com/archive/2005/02/14/050214crci_cinema
50. In a nice post-modern touch, no sooner had the take-down request been reported than a new spoof appeared, in which Hitler is ranting about being taken off YouTube.
51. http://www.jstor.org/pss/764003
52. T.S. Eliot, *The Sacred Wood: Essays on Poetry and Criticism*, Faber & Faber, 1921.
53. Jaron Lanier, 'Web 2.0 is Utterly Pathetic', *Independent*, 10 February 2010. Online at http://www.independent.co.uk/life-style/gadgets-and-tech/features/jaron-lanier-web-20-is-utterly-pathetic-1894257.html (accessed 26 May 2010).
54. Jurgen Habermas, *Structural Transformation of the Public Sphere*, MIT Press, 1989.
55. It's worth saying, however, that this optimistic inference is not shared by Habermas himself. The Internet, he told an interviewer, generates 'a centrifugal force' which 'releases an anarchic wave of highly fragmented circuits of communication that infrequently overlap. Of course, the spontaneous and egalitarian nature of unlimited communication can have subversive effects under authoritarian regimes. But the Web itself does not produce any public spheres. Its structure is not suited to focusing the attention of a dispersed public of citizens who form opinions simultaneously on the same topics and contributions which have been scrutinised and filtered by experts.' It's not clear, though, whether

Habermas is actually familiar with online discourse. For example, when challenged about whether his scepticism applied to phenomena like social networking, he admitted that since he used the Internet 'only for specific purposes and not very intensively', he had 'no experience of social networks like Facebook and cannot speak to the solidarising effect of electronic communication, if there is any.'

See Stuart Jeffries, 'A Rare Interview with Jurgen Habermas', *Financial Times*, 30 April 2010. Available online at http://www.ft.com/cms/s/0/eda3bcd8-5327-11df-813e-00144feab49a.html (accessed 11 May 2010).

56. Brian Ross and Rehab El-Buri, 'Obama's Pastor: God Damn America, US to Blame for 9/11', ABC News, 13 March 2008, http://abcnews.go.com/Blotter/DemocraticDebate/story?id=4443788&page=1 (accessed 8 April 2011).
57. It ran to 38 minutes.
58. It was founded in 2005.
59. http://www.youtube.com/watch?v=zrp-v2tHaDo and http://www.youtube.com/watch?v=pWe7wTVbLUU (accessed 25 May 2010).
60. William Dutton, 'Through the Network of Networks – The Fifth Estate', Inaugural Lecture, University of Oxford, 15 October 2007. Text available online at http://papers.ssrn.com/sol3/papers.cfm?abstract_id=1134502 (accessed 3 June 2010).
61. Kevin Kelly, 'Scan this Book!', *New York Times*

Magazine, 14 May 2006. Available online at http://www.nytimes.com/2006/05/14/magazine/14publishing.html (accessed 25 May 2010).

62. *ibid.*
63. Bob Thompson, 'Explosive Words', *Washington Post*, 22 May 2006, p C01).
64. Jaron Lanier, *You Are Not a Gadget: A Manifesto*, Penguin 2010, p.46.
65. Robert Darnton, 'Extraordinary Commonplaces', *New York Review of Books*, 21 December 2000. Available online at http://www.nybooks.com/articles/archives/2000/dec/21/extraordinary-commonplaces/ (accessed 26 May 2010).
66. Steven Berlin Johnson, 'The Glass Box and the Commonplace Box: two paths for the future of text', Hearst New Media Lecture, Columbia University, April 2010. Available online at http://www.stevenberlin-johnson.com/2010/04/the-glass-box-and-the-common-place-book.html (accessed 26 May 2010).
67. Darnton, *op. cit.*
68. The transformation of cargo shipping by the introduction of the standardised metal container which could be loaded directly onto trucks.
69. Matt Ridley, 'Humans: Why They Triumphed', *Wall Street Journal*, 22 May 2010. Online at http://online.wsj.com/article/SB10001424052748703691804575254533386933138.html (accessed 3 June 2010).
70. Johnson, *op. cit.*
71. The anthropologist Mike Wesch has a nice YouTube

video which illustrates this. http://www.youtube.com/watch?v=6gmP4nk0EOE (accessed 17 April 2011).

5. Complexity is the new reality

1. See Neil Postman, *Technopoly: The Surrender of Culture to Technology*, Vintage Books, 1993, pp. 3–4.
2. By 1556 something resembling a monthly 'newspaper' was being published in Venice.
3. Tom Standage, *The Victorian Internet: The Remarkable Story of the Telegraph and the Nineteenth Century's Online Pioneers*, Weidenfeld and Nicolson, 1998.
4. The *Daily Mirror*, the first newspaper aimed squarely at a popular market (and priced accordingly) was first published in London in 1903.
5. See James Ladyman, James Lambert and Karoline Wiesner, 'What is a complex system?', 2011, online at http://philsci-archive.pitt.edu/8496/ (accessed 15 June 2011).
6. Herbert A. Simon, 'The Architecture of Complexity', *Proceedings of the American Philosophical Society*, Vol. 106, No. 6, 12 December 1962, pp.467–82.
7. See Ladyman,Lambert and Wiesner, *op. cit.*
8. The phrase originates from a discovery by the meteorologist Edward Lorenz, who first discovered the significance that tiny changes in initial conditions can have for a system's subsequent behaviour. In 1961, when using a numerical computer model to rerun a weather prediction, he entered the decimal .506 instead of entering the full .506127 as a shortcut in defining the initial condi-

tions for a simulation run. The result was a completely different weather scenario. Lorenz published his findings in a 1963 paper in which he noted that 'One meteorologist remarked that if the theory were correct, one flap of a seagull's wings could change the course of weather forever.' In later speeches and papers he used the more poetic butterfly. Wikipedia puts it this way: 'The flapping wing represents a small change in the initial condition of the system, which causes a chain of events leading to large-scale alterations of events (compare: domino effect). Had the butterfly not flapped its wings, the trajectory of the system might have been vastly different. While the butterfly does not "cause" the tornado in the sense of providing the energy for the tornado, it does "cause" it in the sense that the flap of its wings is an essential part of the initial conditions resulting in a tornado, and without that flap that particular tornado would not have existed.' (http://en.wikipedia.org/wiki/Butterfly_effect. Accessed 2 January 2010).

9. Just to underscore the point, there is an entire branch of mathematics known as chaos theory.
10. Seth Lloyd, 'Measures of Complexity: a non-exhaustive list', *IEEE Control Systems Magazine*, Vol. 21, No. 4, 2001, pp.7–8.
11. Yochai Benkler, *The Wealth of Networks*, Yale,2006, p.4.
12. *ibid.*
13. http://www.internetworldstats.com/stats.htm (accessed 3 January 2011).

14. http://www.shirky.com/writings/weblogs_publishing.html (accessed 3 January 2011).
15. As of 15 June 2011. See http://twitter.com/#!/stephenfry.
16. Howard Rheingold, *Smart Mobs: The Next Social Revolution*, Persus Books, 2003.
17. Jemima Kiss, 'Facebook began as a geek's hobby. Now it's more popular than Google', *Guardian*, 4 January 2011, p.23. Online at http://www.guardian.co.uk/technology/2011/jan/04/faceboook-mark-zuckerberg-google (accessed 4 January 2011).
18. *ibid.*
19. Jonathan Rosenhead, 'Complexity Theory and Management Practice', http://human-nature.com/science-as-culture/rosenhead.html (accessed 20 October 2010).

6. The network is now the computer

1. I haven't been able to identify a precise date for the quotation, but Gage was the twenty-first employee of Sun Microsystems, a start-up founded in 1982 by engineers from Stanford and the University of California at Berkeley. He joined the company shortly after its foundation and served as its Chief Researcher and Vice President of its Science Office. Sun was distinctive because from the beginning its workstations came with Ethernet local-area networking and TCP/IP (i.e. Internet) connectivity built in. So it was not entirely surprising that Gage's aphorism became the company's official motto.
2. See Michael Riordan, *Crystal Fire: Invention of the*

Transistor and the Birth of the Information Age, W. W. Norton & Co, 1999.

3. The prediction was supposedly made in 1943, but it may be an urban myth. At any rate I have been unable to find any reliable source for it.
4. Nicholas Carr, *The Big Switch: Rewiring the World from Edison to Google*, W.W. Norton, 2009, p.50.
5. See http://en.wikipedia.org/wiki/Semi_Automatic_Ground_Environment (accessed 15 June 2011).
6. Carr, *op cit.*, p.52.
7. James Wallace and Jim Erickson, *Hard Drive: Bill Gates and the Making of the Microsoft Empire*, Wiley, 1992, p.74.
8. The word comes from busbar, a term used in the electricity industry to describe a thick strip of copper or aluminium that conducts electricity between various kinds of electrical apparatus.
9. http://www.searchenginehistory.com/ (accessed 15 October 2010).
10. Carr, *op. cit.*, p.113.
11. Carr, *op. cit*, p.11.
12. Carr, *op. cit*, p.12.
13. John Hagel & John Seely Brown, 'Cloud Computing: Storms on the Horizon', Deloitte Center for the Edge, 2010.
14. John Parales, 'David Bowie, 21st-Century Entrepreneur', *New York Times*, 9 June 2002. Online at http://www.nytimes.com/2002/06/09/arts/david-bowie-

21st-century-entrepreneur.html (accessed 22 November 2010).

15. John Naughton, 'Amazon's new Cloud Drive rains on everyone's parade', *Observer*, 3 April 2011. Online at: http://www.guardian.co.uk/technology/2011/apr/03/john-naughton-amazon-cloud-drive-google-sony (accessed 8 April 2011).
16. For more detailed explication of these models see Chris Anderson, *Free: How Today's Smartest Businesses Profit by Giving Something for Nothing*, Random House, 2010.
17. Dave Winer, 'Welcome to your hamster cage!', 14 November 2010, http://scripting.com/stories/2010/11/14/welcomeToYourHamsterCage.html (accessed 22 November 2010).
18. Nicholas Carr, 'Sharecropping the long tail', 19 December 2006. http://www.roughtype.com/archives/2006/12/sharecropping_t.php (accessed 22 November 2010).
19. Hagel and Seely Brown, *op. cit.*
20. But note that subscription-based charges are different from pay-as-you-use services, in that the customer pays them whether or not they are used.
21. 'Cloudy with a chance of rain', *The Economist*, 5 March 2010. Online at http://www.economist.com/node/15640793 (accessed 22 November 2010).
22. The BBC's Technology Correspondent, Rory Cellan-Jones, recorded an interesting video-tour of one of Microsoft's data centres. See http://news.bbc.co.uk/1/hi/technology/7694471.stm (accessed 8 April 2011).
23. http://royal.pingdom.com/2008/04/11/map-of-all-google-

data-center-locations/ (accessed 23 November 2010).

24. The phrase comes from a sonnet, 'High Flight', written by John Gillsepie Magee, a young Canadian pilot who flew a Spitfire in the Battle of Britain and died in December 1941 at the age of nineteen. The lines were made famous by President Ronald Reagan, who quoted them in his speech to the nation on 28 January 1986 after the Challenger space mission ended in disaster. See http://www.qunl.com/rees0008.html (accessed 23 November 2010).
25. Eric Williams, 'Environmental Impacts in the Production of Personal Computers', in Reudiger Kuehr and Eric Williams (editors), *Computers and the Environment: Understanding and Managing their Impacts*, Kluwer Academic Publishers, 2003, pp.41–72.
26. http://www.gartner.com/it/page.jsp?id=530912 (accessed 23 November 2010).
27. 'Climate Group and the Global e-Sustainability Initiative (GeSI)(2008). SMART 2020: enabling the low carbon economy in the information age'. Available at http://www.smart2020.org/_assets/files/03_Smart2020Report_lo_res.pdf.
28. That might still be an improvement on the present position, in that total ICT-related carbon emissions might decline in a world where more computing is done in large, energy-efficient data centres and less in offices and local networks. A study by Accenture, a consulting company, commissioned by Microsoft, compared energy use in data centres running cloud versions of three key Microsoft enterprise products with energy use in organi-

zations running the same software on their own, premises-based, IT systems. The study found that, for large deployments, Microsoft's cloud service reduced energy use and carbon emissions by more than thirty per cent. The savings were even greater for small deployments with energy use and emissions reduced by more than ninety per cent. See *Cloud Computing and Sustainability: The Environmental Benefits of Moving to the Cloud*, Accenture, 2010. Available online from https://microsite.accenture.com/sustainability/research_and_insights/Documents/Accenture_Sustainability_Cloud_Computing_The%20Environmental%20Benefits%20of%20Moving%20to%20the%20Cloud.pdf.

29. Alex Wissner-Gross, 'How you can help reduce the footprint of the Web', *Sunday Times*, 11 January 2009. http://www.timesonline.co.uk/tol/news/environment/article5488934.ece (accessed 23 November 2010).
30. Jonathan Leake and Richard Woods, 'Revealed: the environmental impact of Google searches', *Sunday Times*, 11 January 2009. http://technology.timesonline.co.uk/tol/news/tech_and_web/article5489134.ece (accessed 23 November 2010).
31. Nicholas Carr, 'Avatars consume as much electricity as Brazilians', Blog post, 5 December 2006. http://www.roughtype.com/archives/2006/12/avatars_consume.php (accessed 23 November 2010).
32. Portable Document Format, a *de facto* standard originally developed by Adobe for distributing documents when it's important to preserve the look and feel of the original.

7. The Web is evolving

1. http://wordnetweb.princeton.edu/
2. http://www.w3.org/People/Berners-Lee/Weaving/glossary.html
3. Unique addresses are provided by the Uniform Resource Locator or URL; mark-up is accomplished by Hypertext Markup Language – HTML; and transport is organized by the Hypertext Transfer Protocol —HTTP.
4. The Official Google Blog, 'We knew the Web was big . . .', 25 July 2008. http://googleblog.blogspot.com/2008/07/we-knew-web-was-big.html (accessed 24 November 2010).
5. You can find up-to-date estimates at http://www.worldwidewebsize.com/.
6. http://www.archive.org/ (accessed 24 November 2010).
7. For a personal account of the origins of the Web see Tim Berners-Lee, *Weaving the Web*, Oxford, 1999.
8. There's a detailed account of the evolution of hypertext in chapter 14 of my book, *A Brief History of the Future: The Origins of the Internet* (Phoenix, 2000).
9. An acronym for Hypertext Markup Language.
10. An acronym for Hypertext Transfer Protocol.
11. http://en.wikipedia.org/wiki/Perl (accessed 29 November 2010).
12. See Rory Cellan-Jones, *Dot Bomb: The Rise and Fall of Dot.com Britain*, Aurum, 2001, and John Cassidy, *Dot.Con: The Greatest Story Ever Sold*, Penguin, 2002.

13. An archived version of the conference website can be found at http://web.archive.org/web/20040602111547/http://web2con.com/ (accessed 7 December 2010).
14. See http://oreilly.com/. The company also organizes high-profile conferences on aspects of information technology.
15. 'The Great Predictor', *Newsweek*, 10 December 2010. http://www.newsweek.com/2010/12/10/the-future-according-to-tim-o-reilly.html (accessed 12 December 2010).
16. Tim O'Reilly, 'What Is Web 2.0? Design Patterns and Business Models for the Next Generation of Software', 30 September 2005. http://oreilly.com/pub/a/web2/archive/what-is-web-20.html (accessed 1 December 2010).
17. *ibid.*
18. The term 'peer production' comes from Yochai Benkler. See his book, *The Wealth of Networks*, Yale, 2006.
19. *ibid.* Emphasis added.
20. Nicholas Carr, 'Sharecropping the long tail', blog post, 19 December 2006. http://www.roughtype.com/archives/2006/12/sharecropping_t.php (accessed 7 December 2010).
21. See http://en.wikipedia.org/wiki/RSS for more detail.
22. See http://www.smallpieces.com/.
23. O'Reilly, *op. cit.*
24. http://en.wikipedia.org/wiki/Software_release_life_cycle (accessed 2 December 2010).

25. In the software business a 'build' is an executable version of a program – i.e. one that can run on its target machine. Most complex programs are written (and revised) in a high-level, human-readable programming language which produces what is called 'source code'. This then has to be translated ('compiled') into low-level machine code that the processor can understand and obey. The end-point of the compilation process is called a build. Development of a complex program can involve tens of thousands of builds. See http://en.wikipedia.org/wiki/Software_build (accessed 17 June 2011).
26. O'Reilly, *op. cit.*
27. Phil Wainewright, 'Why Microsoft can't best Google', 24 August 2005. http://www.zdnet.com/blog/saas/why-microsoft-cant-best-google/13 (accessed 2 December 2010).
28. O'Reilly, *op. cit.*
29. Among the other names that have been canvassed are 'Web Squared' and 'The Social Web', but neither seems to have gained much traction.
30. John Markoff, 'Entrepreneurs see a Web based on Common Sense', *New York Times*, 12 November 2006 (accessed 13 December 2010).
31. See, for example, http://www.guardian.co.uk/world-government-data (accessed 28 December 2010).
32. My colleague Ray Corrigan points out that these kinds of mashups can sometimes be too simplistic and lead to the drawing of erroneous conclusions. For example, in one case the 'crime centre' of Portsmouth appeared to be

a short street with a pub, a car park and a block of flats. But it turned out that the street's postcode was used by police to record retail crime at the local shopping centre and violent crime at neighbouring bars and clubs. The message is that careless or ill-informed use of such data may sometimes be worse than no use at all.

33. http://labs.timesonline.co.uk/blog/2009/03/11/uk-cycling-accidents/ (accessed 28 December 2010).
34. http://www.youtube.com/watch?v=HeUrEh-nqtU (accessed 3 October, 2011).

8. Copyrights and 'copywrongs': or why our intellectual property regime no longer makes sense

1. Marcus Boon, *In Praise of Copying*, Harvard University Press, 2010.
2. Matt Ridley, *The Rational Optimist: How Prosperity Evolves*, Fourth Estate, 2010.
3. Steven Johnson, *Where Good Ideas Come From: The Natural History of Innovation*, Allen Lane, 2010. See also his TED talk http://www.ted.com/talks/steven_johnson_where_good_ideas_come_from.html (accessed 2 February 2011).
4. Johnson, *op. cit.*.
5. Lawrence Lessig, *The Future of Ideas: The Face of the Commons in a Connected World*, Random House, 2001, p.8.
6. For anonymous works, pseudonymous works or works made for hire that have been published since 1978 – which basically means stuff made by corporations like Hollywood studios – the term is 95 years from publica-

tion or 120 years from creation (whichever is the shorter). See http://en.wikipedia.org/wiki/List_of_countries%27_copyright_length (accessed 2 February 2011).

7. Though a trademark can have unlimited duration if the owner continues to use and defend it.
8. Thomas Jefferson, letter to Isaac McPherson, 13 August 1813. Text online at http://press-pubs.uchicago.edu/founders/documents/a1_8_8s12.html (accessed 2 February 2011).
9. Called a 'checksum'. See http://en.wikipedia.org/wiki/Checksum (accessed 2 February 2011).
10. As Jaron Lanier points out in his widely read polemic *You Are Not A Gadget: A Manifesto*, Allen Lane, 2010.
11. For example the film *The Social Network*, which is discussed in Chapter 3, was edited with Final Cut Pro. See http://www.imdb.com/title/tt1285016/trivia (accessed 11 February 2011).
12. http://www.youtube.com/watch?v=rNpoS6jOty4 (accessed 3 February 2011).
13. You can see examples of his work at http://soderberg.tv/ (accessed 2 February 2011).
14. Lawrence Lessig, *Remix: Making Art and Commerce Thrive in the Hybrid Economy*, Bloomsbury Academic, 2008.
15. Peter Yu, *Intellectual Property and Information Wealth: Copyright and Related Rights*, Praeger, 2007, p.34.
16. Paul Henry Lang, *George Frideric Handel*, W.W. Norton, 1966, p.567.

17. J. Peter Burkholder, *Charles Ives and the Uses of Musical Borrowing*, Yale, 2004.
18. Lessig, *op. cit.*, pp.1–5.
19. http://www.eff.org/cases/lenz-v-universal
20. http://www.youtube.com/watch?v=N1KfJHFWlhQ (accessed 9 February 2011).
21. Lessig, *op. cit.*, p.4.
22. Evan Eisenberg, *The Recording Angel: Music, Records and Culture from Aristotle to Zappa*, Yale, 2005.
23. Fulton Brylawski and Abe Goldman, *Legislative History of the 1909 Copyright Act*, Rothman, 1976, p.4.
24. Lessig, *op. cit.*, p.27.
25. Henry Jenkins, *Convergence Culture: Where Old and New Media Collide*, New York University Press, 2006, p.139.
26. Lessig, *op. cit.*, p.29.
27. US Supreme Court, *United States vs. Causby*. Available online at http://supreme.justia.com/us/328/256/case.html (accessed 8 February 2011).
28. Text available online at http://thomas.loc.gov/cgi-bin/query/z?c105:H.R.2281.ENR: (accessed 8 April 2011).

9. Orwell vs. Huxley: the bookends of our networked future?

1. Neil Postman, *Amusing Ourselves to Death: Public Discourse in the Age of Show Business*, Heinemann, 1985, p.155.
2. See his book *Technopoly: The Surrender of Culture to Technology*, Vintage, 1993.

3. In fact, the only record I can find of Postman participating in a public discussion about the subject is in a forum hosted by PBS in 1996 in which he displays a distinctly old-fashioned attitude towards cyberspace. See PBS.org, 'Neil Postman Ponders High Tech', 17 January 1996. http://www.pbs.org/newshour/forum/january96/postman_1-17.html (accessed 5 January 2011).
4. Some of his friends found Orwell's hectoring tone a bit trying. Evelyn Waugh, for example, famously said that Orwell 'could not blow his nose without moralizing about working conditions in the handkerchief industry'.
5. http://en.wikipedia.org/wiki/Orwellian (accessed 10 January 2011).
6. George Orwell, *Nineteen Eighty-Four*, Penguin, 1954. The first edition of the book was published by Secker and Warburg in 1949.
7. Inside your domestic or office network, your computer has a unique address which takes a similar form, but which is not visible to the outside world.
8. See http://en.wikipedia.org/wiki/Anonymous_web_browsing (accessed 11 January 2011).
9. Patrick Kingsley, 'Julian Assange tells students that the web is the greatest spying machine ever', *Guardian*, 15 March 2011. Online at: http://www.guardian.co.uk/media/2011/mar/15/web-spying-machine-julian-assange (accessed 8 April 2011).
10. http://en.wikisource.org/wiki/Communications_Assistance_for_Law_Enforcement_Act (accessed 29 October 2010).

11. The Internet works by dividing messages, pages, music files, etc. into *data packets* which are then transported through the network. Packet sniffers are computer programs that intercept these packets as they pass through an ISP's servers, and hoover them up in order that their contents can be examined using other programs. A packet sniffer is thus an information-gathering tool rather than an analytical tool.
12. http://en.wikipedia.org/wiki/Computer_surveillance (accessed 11 January 2011).
13. http://www.fipr.org/rip/ripa2000.htm (accessed 11 January 2011).
14. Home Office: Business Plan 2011–15, p. 21. Online at http://www.homeoffice.gov.uk/publications/about-us/corporate-publications/business-plan-2011-15/business-plan (accessed 11 January 2011).
15. Nicholas Murray, *Aldous Huxley: An English Intellectual*, Little Brown, 2002, pp.96–109.
16. He was colloquially known as 'Darwin's Bulldog'.
17. Murray, *op. cit.*, p.1.
18. 'O wonder! How many goodly creatures are there here! How beauteous mankind is! O brave new world! That has such people in it!' (*The Tempest*, Act V, Scene 1.)
19. Murray, *op. cit.*, p.250.
20. Aldous Huxley, *Brave New World Revisited*, Triad Books, 1983.
21. Postman, *Amusing Ourselves to Death*, p.151–2.
22. Clay Shirky, 'Weblogs and the Mass Amateurization of

Publishing', http://www.shirky.com/writings/weblogs_publishing.html (accessed 2 January 2011).

23. Bernardo A. Huberman, *The Laws of the Web: Patterns in the Ecology of Information*, MIT Press, 2001.
24. As measured by Alexa.com, a leading web-metrics firm.
25. Not counting Wikipedia, which is run by a not-for-profit organization based in the US.
26. Text of the act is available on the website of the Federal Communications Commission at http://www.fcc.gov/Reports/tcom1996.txt (accessed 12 December 2010).
27. Tim Wu, *The Master Switch: The Rise and Fall of Information Empires*, Alfred Knopf, 2010.
28. Wu, *op. cit.*, p.6.
29. Wu, *op. cit.*, p.7.
30. Wu, *op. cit.*, p.37.
31. Wu, *op. cit.*, p.38.
32 Tim Wu, lecture at the Berkman Center, Harvard University, 11 January 2011. Available online at http://cyber.law.harvard.edu/interactive/events/luncheon/2011/01/wu (accessed 15 January 2011).
33. Farhad Manjoo, 'Why Steve Jobs Won't Return to Apple', *Slate*, 17 January 2011. http://www.slate.com/id/2281453/ (accessed 28 January 2011).
34. http://en.wikipedia.org/wiki/ITunes_Store (accessed 28 January 2011).
35. A new range of iPods with touch-screen interfaces also ran iOS.
36. As were some apps which appeared to compete with

services offered by the mobile networks that were Apple's commercial partners in the phone business. For example in 2009 Apple refused to approve the Google Voice application for iPhone and removed other related third-party applications from its App Store. This prompted an investigation by the Federal Communications Commission.

37. When I last checked (early in 2011), more than 300,000 apps were available for free or paid download.
38. It's not clear whether Google ever intended to become a handset manufacturer, but in the end it has developed only one branded phone of its own – the Nexus. Most handsets running Android are made by other companies.
39. Jonathan Zittrain, *The Future of the Internet – and How to Stop It*, Penguin, 2009. Freely downloadable from http://futureoftheinternet.org/ (accessed 28 January 2011).
40. John Markoff, 'Steve Jobs Walks the Tightrope Again', *New York Times*, 12 January 2007. Online at http://www.nytimes.com/2007/01/12/technology/12apple.html
41. Electronic Frontier Foundation, 'EFF Analysis of the Provisions of the USA PATRIOT Act', 27 October 2003. http://w2.eff.org/privacy/Surveillance/Terrorism/20011031_eff_usa_patriot_analysis.php (accessed 25 January 2011).
42. http://www.fipr.org/rip/ripa2000.htm (accessed 25 January 2011).
43. Paradoxically, it was members of the (unelected) House

of Lords who did the most useful work in amending the bill and making some of its provisions less toxic to civil liberties.

44. Evgeny Morozov, *The Net Delusion: The Dark Side of Internet Freedom*, Public Affairs, 2011.
45. *ibid.*

Appendix

1. Katie Hafner and Matthew Lyon, *Where Wizards Stay Up Late*, Simon & Schuster, 1998, p.12.
2. Hafner and Lyon, *op. cit.*.
3. See Hafner and Lyon, op. cit., pp.62-4 and John Naughton, *A Brief History of the Future: The Origins of the Internet*, Phoenix, 2000, pp.106-8.
4. This idea of not privileging any particular data packet is at the heart of what later became known as 'Net neutrality'.

INDEX

Note: page numbers in **bold** refer to illustrations.

INDEX

INDEX

INDEX

INDEX

INDEX